"双高计划"双语教学教材

Process and Equipment of
Chemical Reaction

化学反应过程与设备

（双语版）

程 进 于 荟 主编

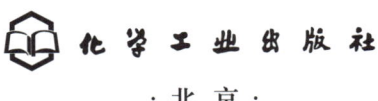

化学工业出版社

·北京·

内容简介

　　《化学反应过程与设备（双语版）》依据化工技术类专业课程标准编写，内容覆盖化学反应器的选择、设计、操作与控制，重点讲解均相反应器、气固相反应器和气液相反应器等基本原理与应用。教材采用双语编写方式，既帮助国际学生深入理解化学反应技术，又为我国学生提供了提升专业英语水平的机会，培养其在全球化职业环境中的竞争力。本教材遵循化学反应器的相关工作流程，强调学生岗位能力的培养。通过"英文知识体系讲解"+"中文核心技术理论讲解"+"中文关键词讲解"+"英文练习巩固"的独特设计，既避免了语言难度与专业知识难度的叠加，又提升了学生的英语阅读能力和专业理解。本书配套动画等数字化学习资源，充分支持线上、线下及混合式教学，增强了教学的互动性与学习效果。

　　本教材适用于高职院校化工技术类及相关专业的学生，也可供化工企业职工的继续教育和在职培训使用，帮助学生和从业人员系统掌握化学反应器的核心技术，为职业生涯打下坚实基础。

图书在版编目（CIP）数据

　　化学反应过程与设备：英、汉 / 程进，于荟主编.
北京：化学工业出版社，2025. 6. -- （"双高计划"双语教学教材）. -- ISBN 978-7-122-47830-6

　　I. TQ052

　　中国国家版本馆 CIP 数据核字第 2025D5S111 号

责任编辑：吕　尤　徐雅妮　　　　　　装帧设计：韩　飞
责任校对：宋　玮

出版发行：化学工业出版社
　　　　　（北京市东城区青年湖南街 13 号　邮政编码 100011）
印　　装：天津千鹤文化传播有限公司
787mm×1092mm　1/16　印张 15　字数 309 千字
2025 年 8 月北京第 1 版第 1 次印刷

购书咨询：010-64518888　　　　　　售后服务：010-64518899
网　　址：http://www.cip.com.cn
凡购买本书，如有缺损质量问题，本社销售中心负责调换。

定　　价：49.00 元　　　　　　　　　版权所有　违者必究

前　言

2020 年，教育部等八部门印发的《关于加快和扩大新时代教育对外开放的意见》明确指出，教育对外开放是教育现代化的重要特征和推动力。《化学反应过程与设备（双语版）》的编写，紧密契合这一国家战略，旨在为职业院校的师生提供一本集国内外前沿技术与职业技能于一体的教材。　通过双语编写，本教材不仅为国际学生提供专业化学习的资源，助力职教体系服务"一带一路"战略，同时也为我国学生提升专业英语水平提供了有力支持，使他们在全球化的职业环境中更加从容应对。

在整体设计上，本教材基于《化学反应过程与设备（第四版）》（陈炳和、许宁主编）框架，融汇国际先进的教育理念与实践案例，遵循"能力为本、产学结合"的原则，重点围绕化学反应器的选择、设计、操作与控制展开，强化学生的动手实践能力和岗位专业技能培养。

在单元设计上，本教材采用了"英文知识体系讲解"＋"中文核心技术理论讲解"＋"中文关键词讲解"＋"英文练习巩固"的结构方式。　这一设计避免了全英文教材常见的语言与专业知识难度叠加导致的"阅读断路"，同时也避免了中英对照双语教材常见的避开英文看中文的"阅读短路"。　通过这种方式，学生不仅能够在巩固专业知识的同时提升英语阅读水平，还能掌握用英文阅读理解知识体系的技巧。　此外，英文练习巩固还便于教学过程中进行课堂互动，帮助学生更好地理解和掌握知识内容。

在资源配套上，本教材通过国家教学资源库、省级精品在线开放课程及化学工业出版社易课堂等优质学习平台，配套了丰富的数字化课程资源，包括微课、动画、测试题等，全面支持双语课堂的线上、线下及混合式教学实施，提升教学互动性与效果，增强学生的学习体验。

本教材由常州工程职业技术学院程进博士（编写任务 1～10，统稿）、南京科技职业学院于荟老师（编写任务 11～20）担任主编，常州工程职业技术学院伍士国博士（编写习题）担任副主编，常州工程职业技术学院陆敏教授主审。　在编写过程中，陈炳和教授、许宁教授、薛叙明教授和巴斯夫前高级经理 Axel Hilderbrandt 博士给予了宝贵的指导与支持。　在此，我们衷心感谢所有参与教材编写的专家、院校以及所有给予支持的单位。

本教材适用于高职院校化工技术类及相关专业的学生，也适合化工企业职工的在职培训与继续教育使用。 通过本教材的学习，学生将掌握化学反应器的基本原理和应用技能，为未来的职业生涯奠定坚实基础。 我们期待通过本教材为职业教育的国际化发展贡献力量。

编　者
2024 年 10 月

目　录

项目二

Selection, Design, Operation and Control of Gas-Solid Phase Reactors
气固相反应器选择、设计、操作与控制

项目三

Selection, Design, Operation and Control of Gas-Liquid Phase Reactors
气液相反应器选择、设计、操作与控制

参考文献

Selection，Design，Operation and Control of Homogeneous Reactors

均相反应器选择、设计、操作与控制

Selection，Design，Operation and Control of Homogeneous Reactors
均相反应器选择、设计、操作与控制

任务 1 Selection of Homogeneous Reactors 均相反应器选择
任务 2 Design of BRs 间歇釜式反应器设计
任务 3 Design of CSTRs连续釜式反应器设计
任务 4 Design and Selection of Supporting Facilities for Tank Reactors 釜式反应器配套设施设计与选择
任务 5 Design of PFRs 连续管式反应器设计
任务 6 Design and Operation Optimization of Homogeneous Reactors 均相反应器设计与操作优化
任务 7 Operation and Control of Atmospheric BRs 常压间歇釜式反应器操作与控制
任务 8 Operation and Control of High-Pressure BRs 高压间歇釜式反应器操作与控制
任务 9 Operation and Control of CSTRs 连续釜式反应器操作与控制
任务 10 Operation and Control of PFRs Reactors 连续管式反应器操作与控制

任务1

Selection of Homogeneous Reactors
均相反应器选择

任务要点

　　本任务旨在系统讲解均相反应器的基础知识，包括其分类、特点、操作方式及材质选择标准。读者将学习如何根据化学反应的特性选择合适的反应器，掌握间歇、连续、半连续操作模式的优缺点，以及不锈钢、搪瓷、铸铁等材质的适用性。通过本任务，读者将熟悉均相反应器在石油化工、精细化工、医药等领域的典型应用场景，理解设备选型对生产效率和经济性的影响。

学习目标

知识目标

（1）掌握釜式反应器的应用、分类及特点。

（2）掌握釜式反应器在不同操作模式下的特性。

（3）熟悉釜式反应器材质和压力分类的意义及应用场景。

（4）了解管式反应器结构及应用场景。

技能目标

（1）能根据生产需求选择合适的均相反应器类型。

（2）能分析不同反应器材料和压力条件对生产的影响。

价值目标

（1）提高设备选择的合理性，优化生产效率。

（2）注重安全性及环保理念在设备选型中的重要性。

1.1 Applications and Classification of Tank Reactors
釜式反应器应用与分类

　　As one of the most commonly used reactors in chemical production，the *tank reactor*（釜式反应器）is widely applied in various industrial processes.

The tank reactor is characterized by its simple structure，easy operation，and versatility（多用途性），making it suitable for many different types of chemical reactions.

There are several ways to classify tank reactors，including by operation mode，material，and operating pressure. According to operation mode，tank reactors can be classified into batch reactors（BR）（间歇操作釜式反应器，简称间歇式反应器），semi-batch reactors（半连续操作搅拌釜式反应器，简称半连续釜式反应器），and continuous stirred-tank reactors（CSTR）（连续操作釜式反应器，简称为连续操作反应器）. BRs are used for small-scale production and are ideal for reactions that require precise control of reaction time and temperature. CSTRs are used for large-scale production and are suitable for reactions that require a continuous supply of reactants. Semi-batch reactors combine features of both BR and CSTR. In semi-batch operation，some reactants are continuously fed into the reactor while others are added in a batch process. This type of reactor is suitable for reactions that require the gradual addition of reactants but do not need a fully continuous feed throughout the process.

Another way to classify tank reactors is by material. Tank reactors can be made of various materials，such as steel（钢制），glass-lined steel（搪瓷），and cast iron（铸铁）. The choice of material depends on the nature of the reaction and the operating conditions.

Finally，tank reactors can also be classified by operating pressure. There are two main types of tank reactors：atmospheric pressure reactors（低压釜）and autoclave（高压釜）. Atmospheric pressure reactors are used for reactions that require low pressure，while autoclaves are used for reactions that require high pressure.

In summary，tank reactors are versatile and widely used in chemical production. Understanding the various classifications of tank reactors is important for selecting the appropriate reactor for a specific chemical reaction.

技术理论

釜式反应器可用于液液均相反应，也可用于非均相反应，如非均相液相、液固相、气液相、气液固相等。

釜式反应器按操作方式分类为间歇（分批）式、半连续（半间歇）式和连续式操作（如图 1-1 所示），按材质分为钢制反应釜、搪瓷反应釜及铸铁反应釜，按反应釜所能承受的操作压力可分为低压釜和高压釜。

动画

间歇釜
半间歇釜
连续釜
多釜串联

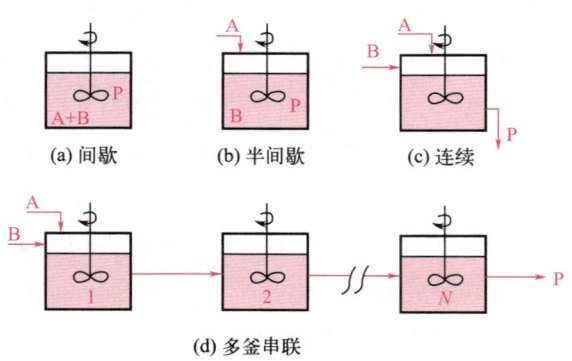

图 1-1 反应釜的操作方式

关键词详解

釜式反应器，tank reactor，普遍应用于石油化工、橡胶、农药、染料、医药等工业，用来完成磺化、硝化、氢化、烃化、聚合、缩合等工艺过程，例如有机染料和医药中间体的生产过程。其特征：结构简单、加工方便、传热传质效率高、操作灵活性大、便于更换品种、能适应多样化的生产。

均相，homogeneous，物料达到分子级别的混合，如食盐水。

非均相，heterogeneous，物料未达到分子级别的混合，如奶茶。

间歇操作，batch operation，物料一次性加入、再一次性取出的操作，通常用于反应时间较长，小批量、多品种的生产，在染料及制药工业中广泛采用这种操作。

半间歇操作，semi-batch operation，部分物料一次性加入，部分物料连续加入的操作，其特别适用于要求一种反应物的浓度高而另一种反应物的浓度低的化学反应，适用于可以通过调节加料速度来控制目标反应温度的反应。

连续式操作，continuous operation，连续操作设备利用率高、产品质量稳定、易于自动控制，适用于大规模生产。

钢制反应釜，steel reactor，其制造工艺简单、造价费用较低、维护检修方便、使用范围广泛。用 Q235A 材料制作的反应釜不耐酸性介质腐蚀，不锈钢材料制的反应釜可以耐一般酸性介质，经过镜面抛光的不锈钢制反应釜特别适用于高黏度体系聚合反应。

搪瓷反应釜，glass-lined steel reactor，许多介质具有良好的抗腐蚀性，所以广泛用于精细化工生产中的卤化反应及有盐酸、硫酸、硝酸等存在时的各种反应。

铸铁反应釜，cast iron reactor，在氯化、磺化、硝化、缩合、硫酸增浓等反应过程中使用较多。

低压釜，atmospheric pressure reactors，最常见的搅拌釜式反应器，在搅拌轴和壳体之间采用动密封结构，在低压（1.6MPa 以下）条件下能够防止物料的泄漏。

高压釜，autoclave，通常采用磁力搅拌釜的静密封结构，实现整台反应器在全密封的转台下工作，确保无泄漏。适用范围：各种剧毒、易燃、易爆以及其他渗透力极强的化工工艺过程，适用于石油化工、有机合成、化学制药、食品等工艺中进行硫化、氟化、氢化、氧化等反应。

▶ 动画

磁力搅拌釜

互动练习

1-1　Which type of reactor is used for large-scale production?

A）BR

B）CSTR

C）Glass-lined steel reactor

D）Hastelloy reactor

1-2　What is the advantage of BRs?

A）They are suitable for large-scale production

B）They require a continuous supply of reactants

C）They allow for precise control of reaction time and temperature

D）They are made of various materials

1-3　What is the primary consideration when choosing the material for a tank reactor?

A）The nature of the reaction

B）The cost of the material

C）The availability of the material

D）The location of the reactor

1-4　What is the main difference between atmospheric pressure reactors and autoclave?

A）The materials used to make them

B）The operating mode

C）The operating pressure

D）The size of the reactor

1-5　What is a disadvantage of tank reactors?

A）They are difficult to operate

B）They require a continuous supply of reactants

C）They are only suitable for specific chemical reactions

D）They can be expensive to build

1.2 Construction of Tank Reactors
釜式反应器结构

The tank reactor is an essential equipment in chemical production. The basic structure of a tank reactor consists of four main parts：the shell（壳），stirring device（搅拌装置），shaft seal（轴封），and heat exchange device（换热装置）.

The shell is the main body of the tank reactor and is usually made of stainless steel or glass-lined steel. The stirring device is used to mix the reactants（反应物）and promote the reaction. The shaft seal is designed to prevent leakage of the reaction mixture，and the heat exchange device is used to regulate the temperature of the reaction.

Another type of tank reactor is the non-leakage magnetic tank reactor（无泄漏磁力釜），which has a more advanced structure. It consists of the kettle body，stirring rotor，heat transfer components，transmission device，and safety and protection devices（安全与保护装置）. The stirring rotor is driven by a magnetic force，which eliminates the need for a mechanical seal and reduces the risk of leakage. The heat transfer components are designed to ensure uniform heating and cooling of the reaction mixture.

Tank reactors have several characteristics and development trends in chemical production. They are widely used for a variety of chemical reactions due to their versatility and ease of operation. The development of the chemical industry has put forward higher requirements for the design and operation of tank reactors，including improving the efficiency of the reaction，reducing energy consumption，and enhancing safety and environmental protection. Therefore，the future development trend of tank reactors will focus on automation（自动化），intelligence（智能化），and integration（集成化），to improve the efficiency and quality of chemical production.

技术理论

釜式反应器主要由釜体（含釜体上连接的压料管、支座及人孔）、搅拌装置（含搅拌器、搅拌轴及传动装置）、轴封和换热装置（如夹套）四大部分组成（如图 1-2 所示）。

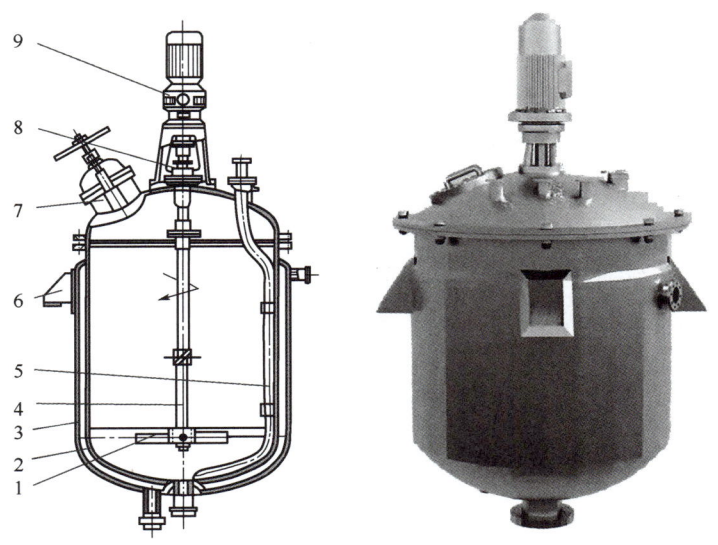

动画

釜式反应器
结构与间歇
操作

图 1-2 釜式反应器的基本结构

1—搅拌器；2—釜体；3—夹套；4—搅拌轴；5—压料管；

6—支座；7—人孔；8—轴封；9—传动装置

无泄漏磁力釜由釜体、搅拌转子、传热构件、传动装置以及安全与保护装置组成。搅拌转子的旋转运动通过磁力驱动器来实现。

关键词详解

釜体，shell，由圆形筒体、上盖、下封头构成。

搅拌装置，stirring device，由搅拌轴和搅拌电机组成，其目的是加强反应釜内物料的混合，以强化反应的传质和传热。

轴封，shaft seal，用来防止釜的主体与搅拌轴之间的泄漏。

换热装置，heat exchange device，用来加热或冷却反应物料，使之符合工艺要求的温度条件。

无泄漏磁力釜，non-leakage magnetic tank reactor，主要用于高压操作或者物料为高毒性、易燃易爆的场合。

互动练习

1-6　What is the main function of the stirring device in a tank reactor?

A）To regulate the temperature of the reaction

B）To prevent leakage of the reaction mixture

C）To promote the reaction

D）To ensure uniform heating and cooling of the reaction mixture

1-7　What is the difference between a conventional tank reactor and a non-leakage magnetic tank reactor?

A）The material used for the shell

B）The operating pressure

C）The type of stirring device

D）The type of shaft seal

1-8　What is the primary advantage of a non-leakage magnetic tank reactor?

A）Lower operating cost

B）Reduced risk of leakage

C）Greater versatility

D）Better temperature control

1-9　What are some development trends for tank reactors in the chemical industry?

A）Increased use of glass-lined steel

B）Greater focus on environmental impact

C）Automation and integration

D）Use of hazardous materials

1-10　What is the main advantage of using a tank reactor in chemical production?

A）Versatility and ease of operation

B）High production volume

C）Lower operating cost

D）Precise control of reaction time

1.3　Applications and Classification of Tubular Reactors 管式反应器应用与分类

The tubular reactor（管式反应器），also known as plug flow reactor（PFR）（活塞流反应器）is an important type of chemical reactor in industrial production. It can be divided into two categories：the multi-tube sequential PFR（多管串联管式反应器）and the multi-tube parallel PFR（多管并联管式反应器）.

The multi-tube sequential PFR is characterized by the reaction taking place sequentially in each tube，which allows for a more controlled reaction process. This type of reactor is often used in industries where precise control of reaction parameters is critical，such as in the production of pharmaceuticals（制药）or specialty chemicals.

The multi-tube parallel PFR is characterized by multiple tubes arranged in parallel，which allows for a higher production rate. This type of reactor is often used in industries where large-scale production is required，such as in the production of bulk chemicals（大型化工）.

The PFR has several unique features that make it suitable for various chemical reactions. For example，the reaction can take place at a high temperature and pressure，which can improve the reaction rate（反应速率）and selectivity（选择性）. Additionally，the PFR has a high heat transfer rate，which can ensure efficient heat exchange and temperature control during the reaction process.

Furthermore，the PFR has a compact structure，which allows for easy integration into a continuous production line. It also has a low level of impurities（杂质）and is easy to clean，which is essential in the production of high-purity（高纯度）products.

In conclusion，the PFR has a wide range of applications in chemical production，and its various types provide options for different production needs. The unique features of the tubular reactor make it an essential tool in the chemical industry，particularly in the production of high-value（高值）products.

技术理论

通常按管式反应器管道的连接方式不同，把管式反应器分为多管串联管式反应器和多管并联管式反应器。管式反应器返混小，因而容积效率（单位容积生产能力）高，对要求转化率较高或有串联副反应的场合尤为适用。此外，管式反应器可实现分段温度控制。其主要缺点是，当反应速率很低时所需管道过长，工业上不易实现。

关键词详解

多管串联管式反应器，multi-tube sequential PFR，结构如图 1-3 所示，一般用于气相反应和气液相反应，例如烃类裂解反应和乙烯液相氧化制乙醛。

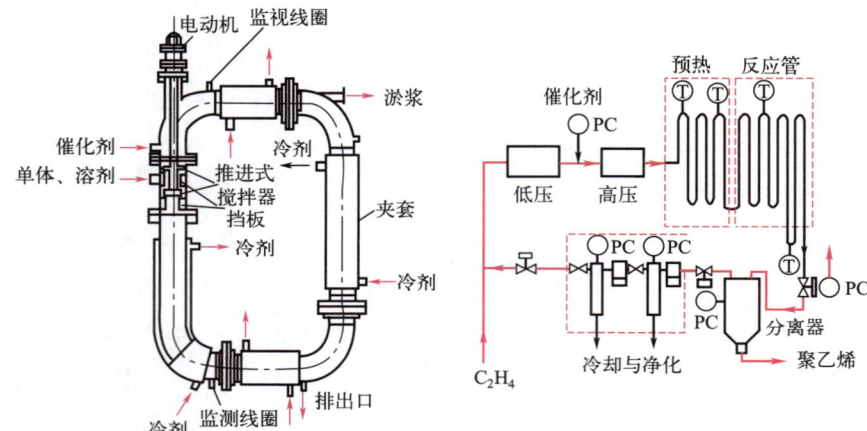

(a) 管式反应器生产高压聚乙烯 　　　　(b) 环管式聚合反应器

图 1-3　多管串联管式反应器结构

多管并联管式反应器，multi-tube parallel PFR，结构如图 1-4 所示，一般用于气固相反应，例如气相氯化氢和乙炔反应制氯乙烯、气相氮和氢制合成氨。

互动练习

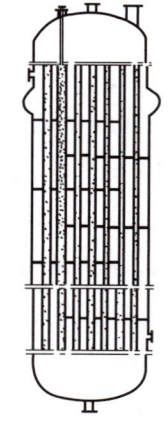

1-11　What are the two categories of PFR?

A）Multi-tube sequential and multi-tube parallel

B）Single-tube and multi-tube

C）Continuous flow and batch

D）High pressure and low pressure

1-12　In which industries is the multi-tube sequential PFR typically used?

图 1-4　多管并联管式反应器结构

A）Bulk chemical production

B）Pharmaceutical production

C）Food production

D）Oil and gas production

1-13　What is the primary advantage of a multi-tube parallel PFR?

A）Precise control of reaction parameters

B）High production rate

C）Low operating cost

D）Easy integration into a continuous production line

1-14　What is the primary advantage of a PFR over other types of reactors?

A）High pressure and temperature tolerance

B）Compact structure

C）Easy to clean

D）All of the above

1-15　What is the importance of a high heat transfer rate in a PFR?

A）It improves the reaction rate and selectivity

B）It allows for precise control of reaction parameters

C）It ensures efficient heat exchange and temperature control

D）It increases the production rate

1.4 Construction of PFRs
管式反应器结构

PFRs are designed with various components that make up their structure. The basic structure consists of a tube（管），which can either be straight or bent，and is typically made of metal or glass. The tube is connected to other components，such as flanges（法兰）and fittings（组件），using sealing rings（密封环）and bolts to prevent leakage of reactants and products.

One important component of the PFR is the thermal compensator（热补偿器），which helps to compensate for thermal expansion（膨胀）and contraction（收缩）during operation. This component is typically made of a flexible material，such as metal bellows（金属波纹管），that can absorb the expansion and contraction of the tube.

Another key component is the heat transfer jacket（热传递夹套），which surrounds the reactor tube and allows efficient heat transfer between the reactor and the cooling or heating medium. The jacket is often made of a material with high thermal conductivity，such as copper or stainless steel.

In addition to these components，the reactor may also include a support frame（支架）or stand to hold the tube in place during operation. This frame may be adjustable to allow for changes in the length of the reactor tube due to thermal expansion.

Overall，the design of the PFR structure is critical to its efficient and safe operation. Properly selecting and installing of components such as sealing rings， flanges，fittings，and heat transfer jackets are essential to ensure that the reactor can withstand the high temperatures，pressures，and corrosive envi-ronments often encountered in chemical processes.

技术理论

以套管式为例，套管式反应器由长径比较大的细长管和密封环组成，两者通过连接件的紧固串联安放在机架上。其包括直管、弯管、密封环（含透镜环管）、管件（含法兰、连接管、螺母、弹性螺柱、连接管、补偿器和支管等）、机架等几部分。如图 1-5 所示。

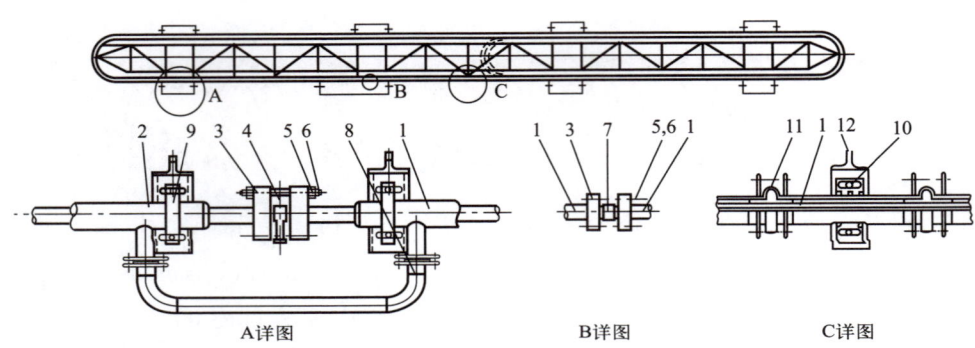

图 1-5 套管式反应器结构

1—直管；2—弯管；3—法兰；4—带接管的"T"形透镜环；5—螺母；6—弹性螺柱；
7—圆柱形透镜环；8—连接管；9—支座（抱箱）；10—支座；11—补偿器；12—机架

关键词详解

直管，straight pipe，由夹套管和内管焊接在一起构成，夹套管上加装补偿器消除焊缝上的拉应力。

弯管，elbow，弯管结构与直管基本相同，弯管在机架上的安装方法允许其有足够的伸缩量，故不再另加补偿器。

密封环，sealing ring，套管式反应器的密封环为透镜环，透镜环有两种形状，一种是圆柱形，另一种是带接管"T"形。圆柱形透镜环采用与反应器内管同一材质制成。带接管的"T"形透镜环用于安装测温、测压元件。

管件，pipe fitting，反应器的连接必须按规定的紧固力矩进行，所以对法兰、螺柱和螺母都有一定要求。

机架，rack，反应器机架用桥梁钢焊接成整体，地脚螺栓安放在基础桩的柱头上，安装管子支座部位装有托架，管子用抱箍与托架固定。

互动练习

1-16   What is the purpose of the thermal compensator in a PFR?

A）To prevent leakage of reactants and products

B）To compensate for thermal expansion and contraction during operation

C）To facilitate efficient heat transfer

D）To hold the tube in place during operation

1-17   Which material is often used for the heat transfer jacket in a PFR?

A）Glass

B）Aluminum

C）Copper

D）Plastic

1-18   What is the function of the support frame in a PFR?

A）To prevent leakage of reactants and products

B）To compensate for thermal expansion and contraction during operation

C）To facilitate efficient heat transfer

D）To hold the tube in place during operation

1-19   Why is the selection of components such as sealing rings and flanges important in the design of a PFR?

A）To prevent thermal expansion and contraction

B）To facilitate efficient heat transfer

C）To prevent leakage of reactants and products

D）To hold the tube in place during operation

1-20   Which of the following is NOT a common material for the thermal compensator in a PFR?

A）Metal bellows

B）Rubber

C）Plastic

D）Fabric

Design of BRs
间歇釜式反应器设计

任务要点

间歇釜式反应器的设计是化工生产的核心环节之一。本任务重点介绍反应器的流动模型（理想置换流动、理想混合流动和非理想流动）以及均相反应动力学方程的应用。读者将学习如何设计适合工业生产的反应器，并考虑操作安全性与经济性，为化工产品生产提供科学依据。

学习目标

知识目标

（1）熟悉间歇釜式反应器的基本设计原理。

（2）掌握反应器流动模型（理想置换流动、理想混合流动及非理想流动）的基础知识。

（3）掌握均相反应动力学基本方程及其应用。

技能目标

（1）能计算反应器体积、直径和高度。

（2）能设计适合工业应用的间歇釜式反应器。

价值目标

（1）在设计中考虑操作安全性和经济性。

（2）强化创新设计能力，适应现代化工需求。

2.1 Chemical Reactor Flow Models
反应器流动模型

Reactors are often modeled using fluid flow models（流体流动模型），which help predict and optimize the performance of a reactor system. Ideal flow models（理想流动模型）assume perfect mixing（充分搅拌），uniform concentrations（均一浓度），and complete plug flow（完全活塞流），among other idealized conditions. Ideal displacement flow models（理想置换流动模型）involve the movement of fresh reactants through the reactor，while spent reactants are displaced out of the reactor. Ideal mixed flow models（理想混合流动模型）involve a balance between plug flow and perfect mixing，where reactants are well-mixed but also flow through the reactor like a plug.

However，flow in a reactor is often non-ideal（非理想）due to factors such as fluid viscosity（黏度），the presence of baffles（挡板）or obstacles，and the rate of mixing. Nonideal flow can lead to issues such as incomplete mixing（不充分搅拌）and the formation of hotspots（热区），which can negatively impact（负面影响）the reaction rate and product quality. Additionally，non-ideal flow can affect the residence time distribution（停留时间分布），which can impact the selectivity of a reaction.

Overall，understanding and modeling the flow behavior in a reactor is important for optimizing reactor performance and achieving the desired reaction outcomes. By considering factors such as ideal and non-ideal flow，researchers and engineers can design and operate reactors to improve reaction efficiency and yield.

技术理论

理想流动模型包括两种：理想置换流动模型和理想混合流动模型。长径比值较大和流量较高的连续管式反应器中的流体流动可视为理想置换流动。搅拌十分强烈的连续釜式反应器中的流体流动可视为理想混合流动。

实际工业反应器中的反应物料流动模型与理想流动有所偏离，往往介于两者之间。对于所有偏离理想置换和理想混合的流动模式统称为非理想流动。

不同时刻进入反应器的物料之间的混合叫做返混。间歇操作反应器中不存在不同时刻进入反应器的物料，理想置换反应器不存在混合，因此前两者返混为零。而理想混合反应器则是返混达到极大。

降低返混程度的主要措施有横向分割和纵向分割两种。

关键词详解

理想置换流动模型，ideal displacement flow model，也称为平推流模型或活塞流模型，如图 2-1 所示，所有的物料在流动的过程中自身发生变化，但沿着流动方向的前后物料互不干扰。

理想混合流动模型，ideal mixed flow model，也称为全混流模型，如图 2-2

所示。由于强烈搅拌，不同时刻进入反应器内的所有物料完全混合。

动画

理想置换流
动模型
理想混合流
动模型

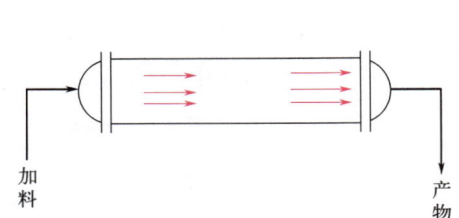

图 2-1　理想置换流动模型

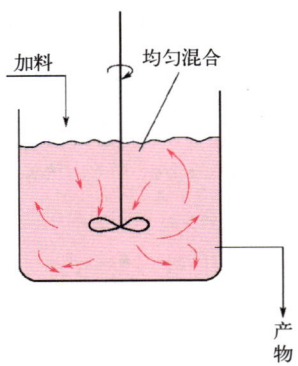

图 2-2　理想混合流动模型

非理想流动，non-ideal flow，由于滞留区、沟流与短路、循环流、流体流速分布不均匀、扩散等多因素的存在，实际流体流动会偏离理想置换流动和理想混合流动模式。

返混，backmixing，其可以用来定义一个非理想流动偏离理想置换流动模型以及理想混合流动模型的程度。

横向分割，horizontal division，如把单个连续釜式反应器改用多釜串联、在流化床反应器内部加装横向挡板、在气液鼓泡反应器中放置填料（填料鼓泡塔）。

纵向分割，vertical division，如在流化床反应器内部设置垂直管作为内部构件、在鼓泡塔反应器中设置垂直管。

互动练习

2-1　What is the purpose of fluid flow models in reactors?

A）To simulate perfect mixing conditions

B）To predict and optimize reactor performance

C）To reduce reactor operating costs

D）To eliminate non-ideal flow behavior

2-2　What are ideal flow models based on?

A）Non-ideal mixing

B）Perfect mixing and complete plug flow

C）Viscosity and obstacles in the reactor

D）Hotspots and reaction rate

2-3　What are some factors that can lead to non-ideal flow in a reactor?

A）Ideal mixing conditions and balanced flow

B）Low viscosity and no obstacles

C）Baffles or obstacles and fluid viscosity

D）High selectivity and residence time distribution

2-4　What is one potential consequence of non-ideal flow in a reactor?

A）Increased reaction efficiency

B）Formation of hotspots

C）Elimination of byproducts

D）Reduced reactor operating costs

2-5　How can understanding flow behavior in a reactor be useful?

A）It can help optimize reactor performance and achieve desired reaction outcomes

B）It can eliminate the need for flow models

C）It can reduce reactor safety risks

D）It can simplify reactor design and operation

2.2 Fundamentals of Homogeneous Reaction Kinetics 均相反应动力学基础

（1）Chemical Reaction Rate and Kinetics Equation

Chemical reaction rate is the measure of how fast a chemical reaction takes place. It is the change in concentration of reactants or products per unit time. The reaction rate can be influenced by several factors such as temperature，pressure，concentration，surface area，and catalysts.

The chemical reaction rate can be expressed mathematically using the chemical reaction kinetics equation（化学反应动力学方程）. The kinetics equation describes the relationship between the reaction rate and the concentrations of reactants and products（反应物与产物浓度）. The most commonly used kinetics equation is the rate law（反应速率方程），which relates the rate of a reaction to the concentrations of reactants.

The rate law for a chemical reaction is determined experimentally and can be expressed as

$$\text{Rate} = k c_A^m c_B^n$$

where k is the rate constant（反应速率常数），A and B are the reactants，and m and n are the reaction orders（反应级数）with respect to A and B，respectively. The reaction order represents how the concentration of each reactant affects the reaction rate.

Understanding the chemical reaction rate and kinetics equation is essential in designing and optimizing chemical reactions for industrial processes. By manipulating the reaction conditions，such as temperature，pressure，and catalysts，it is possible to increase the reaction rate and improve the efficiency of the process.

（2） Homogeneous Reaction Rate and Reaction Kinetics

Homogeneous reaction rate（均相反应速率）and reaction kinetics are important concepts in chemistry. Homogeneous reactions involve only one phase and the reaction rate is expressed in terms of concentration or conversion rate（转化率）.

The reaction rate is dependent on the reactant concentrations and the order of the reaction. The order of a reaction is the sum of the powers to which the concentrations of the reactants are raised in the rate equation. The reaction rate constant is a measure of the rate at which a reaction occurs and is specific to each reaction.

The activation energy（活化能）is the minimum amount of energy required to initiate a reaction. The higher the activation energy，the slower the reaction rate. The Arrhenius equation（阿伦尼乌斯方程）relates the rate constant to the activation energy and the temperature.

Homogeneous single reaction kinetics equations describe the rate of a reaction at constant temperature（恒温）and constant volume（恒容）or constant pressure（恒压）. Complex reaction kinetics equations involve reversible reactions（可逆反应），parallel reactions（平行反应），sequential reactions（串联反应），and complex composite reactions（复合复杂反应）.

In conclusion，understanding homogeneous reaction rate and reaction kinetics is essential in determining the rate of chemical reactions and optimizing reaction conditions for industrial applications.

技术理论

化学反应速率是指在反应系统中，某一物质在单位时间、单位反应区域内的反应量。如式（2-1）所示。均相反应过程的反应区域通常取反应混合物总体积，反应速率单位以 $kmol/(m^3 \cdot h)$ 表示。

$$反应速率 = \frac{反应量}{反应区域 \times 反应时间} \tag{2-1}$$

在恒温时，化学反应动力学方程一般可写成

$$\pm r_i = k f(c_A, c_B, \cdots) \tag{2-2}$$

式中，r_i 为组分 i 的反应速率；c_A，c_B，$\cdots$ 为 A、B、$\cdots$ 组分的浓度；k 为反应速率常数，可由阿伦尼乌斯方程求得。

均相反应的化学反应速率可用以下方法表示。

（1）用组分转化率表示

$$(-r_A) = \frac{n_{A0}}{V} \frac{\mathrm{d}x_A}{\mathrm{d}\tau} \tag{2-3}$$

式中，V 为反应体积，m^3；τ 为反应时间，h。

（2）用浓度表示

$$(-r_A) = -\frac{\mathrm{d}c_A}{\mathrm{d}\tau} \tag{2-4}$$

对于 $a\mathrm{A} + b\mathrm{B} \longrightarrow r\mathrm{R} + s\mathrm{S}$ 反应，根据化学反应计量学可知，各组分的变化量符合下列关系：

$$\frac{n_{A0} - n_A}{a} = \frac{n_{B0} - n_B}{b} = \frac{n_R - n_{R0}}{r} = \frac{n_S - n_{S0}}{s} \tag{2-5}$$

式中，n_{A0}、n_{B0}、n_{R0}、n_{S0} 分别为反应开始时组分的摩尔数；n_A、n_B、n_R、n_S 分别为反应到某一时刻组分的摩尔数。

各组分的反应速率必然满足式（2-6）：

$$\frac{(-r_A)}{a} = \frac{(-r_B)}{b} = \frac{r_R}{r} = \frac{r_S}{s} \tag{2-6}$$

研究反应动力学需要将不同的反应进行分类。

按反应过程涉及几个化学反应式和动力学方程式来表示，可分为单一反应和复杂反应，复杂反应由若干单一反应组成，通常可分为如下几种类型：可逆反应、平行反应、连串反应以及复合复杂反应。

按分子在反应过程中需要相互碰撞的次数来表示，可分为基元反应和非基元反应。其中基元反应按照反应分子数的个数来表示，可以分为单分子反应、双分子反应等。

反应速率对反应物浓度的敏感程度又可以用反应级数来表示。反应级数是指，在均相反应系统中进行不可逆化学反应时，所有浓度/压力项的指数之和，其动力学方程一般都可用式(2-7) 表示，反应级数 $n = \alpha_1 + \alpha_2$。

$$a\mathrm{A} + b\mathrm{B} \longrightarrow r\mathrm{R} + s\mathrm{S}$$

$$\pm r_i = k_i c_\mathrm{A}^{\alpha_1} c_\mathrm{B}^{\alpha_2} \tag{2-7}$$

对于气相反应，也常常使用分压来表示：

$$(-r_\mathrm{A}) = -\frac{1}{V}\frac{\mathrm{d}n_\mathrm{A}}{\mathrm{d}\tau} = k_p p_\mathrm{A}^{\alpha_1} p_\mathrm{B}^{\alpha_2} \tag{2-8}$$

式中，k_p 为以分压表示的反应速率常数。

恒温恒容和恒温变容反应速率方程及其积分形式见表 2-1 和表 2-2。

表 2-1　恒温恒容不可逆反应速率方程及其积分形式

化学反应	速率方程	积分形式
A→P （零级）	$(-r_\mathrm{A}) = -\dfrac{\mathrm{d}c_\mathrm{A}}{\mathrm{d}\tau} = k$	$k\tau = c_{\mathrm{A}0} - c_\mathrm{A} = c_{\mathrm{A}0}x_\mathrm{A}$
A→P （一级）	$(-r_\mathrm{A}) = -\dfrac{\mathrm{d}c_\mathrm{A}}{\mathrm{d}\tau} = kc_\mathrm{A}$	$k\tau = \ln\dfrac{c_{\mathrm{A}0}}{c_\mathrm{A}} = \ln\dfrac{1}{1-x_\mathrm{A}}$
2A→P A+B→P $(c_{\mathrm{A}0}=c_{\mathrm{B}0})$（二级）	$(-r_\mathrm{A}) = -\dfrac{\mathrm{d}c_\mathrm{A}}{\mathrm{d}\tau} = kc_\mathrm{A}^2$	$k\tau = \dfrac{1}{c_\mathrm{A}} - \dfrac{1}{c_{\mathrm{A}0}} = \dfrac{1}{c_{\mathrm{A}0}} \times \left(\dfrac{x_\mathrm{A}}{1-x_\mathrm{A}}\right)$
A+B→P $(c_{\mathrm{A}0} \neq c_{\mathrm{B}0})$（二级）	$(-r_\mathrm{A}) = -\dfrac{\mathrm{d}c_\mathrm{A}}{\mathrm{d}\tau} = kc_\mathrm{A}c_\mathrm{B}$	$k\tau = \dfrac{1}{c_{\mathrm{B}0}-c_{\mathrm{A}0}}\ln\dfrac{c_\mathrm{B}c_{\mathrm{A}0}}{c_\mathrm{A}c_{\mathrm{B}0}} = \dfrac{1}{c_{\mathrm{B}0}-c_{\mathrm{A}0}}\ln\dfrac{1-x_\mathrm{B}}{1-x_\mathrm{A}}$
A→P （n 级）	$(-r_\mathrm{A}) = -\dfrac{\mathrm{d}c_\mathrm{A}}{\mathrm{d}\tau} = kc_\mathrm{A}^n$	$k\tau = \dfrac{1}{n-1}(c_\mathrm{A}^{1-n} - c_{\mathrm{A}0}^{1-n})$ $= \dfrac{1}{c_{\mathrm{A}0}^{n-1}(n-1)}[(1-x_\mathrm{A})^{1-n} - 1]$

表 2-2　恒温变容不可逆反应速率方程的积分式

化学反应	速率方程	积分形式
A→P （零级）	$(-r_\mathrm{A}) = k$	$k\tau = \dfrac{c_{\mathrm{A}0}}{y_{\mathrm{A}0}\varepsilon_\mathrm{A}}\ln(1 + y_{\mathrm{A}0}\delta_\mathrm{A}x_\mathrm{A})$
A→P （一级）	$(-r_\mathrm{A}) = kc_\mathrm{A}$	$k\tau = -\ln(1-x_\mathrm{A})$
2A→P A+B→P $(c_{\mathrm{A}0}=c_{\mathrm{B}0})$（二级）	$(-r_\mathrm{A}) = kc_\mathrm{A}^2$	$c_{\mathrm{A}0}k\tau = \dfrac{(1+\delta_\mathrm{A}y_{\mathrm{A}0})x_\mathrm{A}}{1-x_\mathrm{A}} + \delta_\mathrm{A}y_{\mathrm{A}0}\ln(1-x_\mathrm{A})$

关键词详解

化学反应速率，chemical reaction rate，常以符号 $\pm r_i$ 表示。如果是反应物，在反应速率前赋予负号，如 $-r_A$ 表示反应物 A 的消耗速率。如果是产物，反应速率取正号，如 r_R 表示产物 R 的生成速率。

化学反应动力学方程，chemical reaction kinetics equation，定量描述反应速率与影响反应速率因素之间的关系。

反应速率常数，reaction rate constant，直接决定了反应速率的高低和反应进行的难易程度。其随温度、压力、催化剂的变化而变化。

阿伦尼乌斯方程，Arrhenius equation，如式（2-9）

$$k = A_0 \exp\left(-\frac{E}{RT}\right) \tag{2-9}$$

式中，A_0 为指前因子；E 为反应活化能；R 为气体通用常数，$R = 8.314 \text{kJ}/(\text{kmol} \cdot \text{K})$。

反应活化能，reaction activation energy，其物理含义是指使反应物分子达到活化态所需的能量。反应活化能是反应速率对反应温度敏感程度的一种度量。图 2-3 表明了吸热反应和放热反应中反应活化能和活化态的示意图。

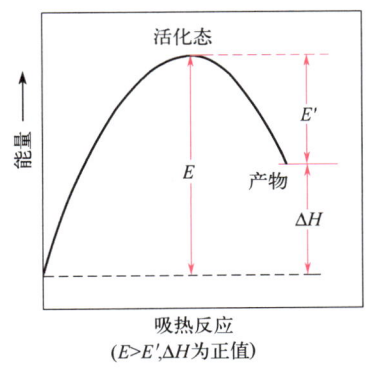
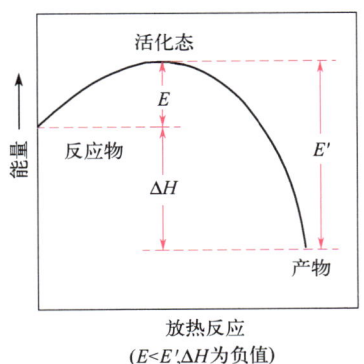

图 2-3 吸热反应和放热反应的能量示意图

转化率，conversions，用 x_A 表示，$x_A = 1 - n_A/n_{A0}$。

单一反应，single reaction，只用一个化学反应式和一个动力学方程式便能加以描述的反应。

复杂反应，complex reaction，有几个反应同时进行，要用几个动力学方程式才能加以描述的反应。常见的复杂反应有：可逆反应、平行反应、连串反应、复合复杂反应等。

可逆反应，reversible reaction，在反应物发生化学反应生成产物的同时，产物之间也在发生化学反应恢复成原料。如：

$$A+B \rightleftharpoons R+S$$

基元反应，elementary reaction，反应物分子在碰撞中一步直接转化为产物分子的反应。

非基元反应，non-elementary reaction，反应物分子要经过若干步，即经由几个基元反应才能转化成为产物分子的反应。

单分子反应，monomolecular reaction，参加反应的分子数为1的反应。

双分子反应，di-molecular reaction，由两个分子碰撞接触的反应。

反应级数，reaction order，是指动力学方程式中浓度项的指数，是由实验确定的常数，反映了反应速率对反应物浓度的敏感程度。

互动练习

2-6　What is the definition of chemical reaction rate?

A）The number of reactants used up in a reaction

B）The time it takes for a reaction to occur

C）The speed at which reactants are converted to products

D）The concentration of products formed in a reaction

2-7　Which of the following is an example of a chemical reaction rate expression?

A）$c_A + c_B \rightarrow c_C$

B）$rate = k\, c_A^2 c_B$

C）$E+S \rightarrow ES \rightarrow E+P$

D）$A+B \rightarrow AB$

2-8　What does the rate constant represent in the chemical reaction rate equation?

A）The overall rate of the reaction

B）The concentration of the reactants

C）The activation energy of the reaction

D）The rate of the reaction at unit concentration of reactants

2-9　Which of the following affects the chemical reaction rate?

A）Temperature

B）Concentration of reactants

C）Catalysts

D）All of the above

2-10　Which of the following is NOT a type of chemical reaction rate equation?

A）Zero-order

B）First-order

C）Second-order

D）Third-order

2-11　Homogeneous reactions involve ＿＿ phase（s）.

A）one

B）two

C）three

D）four

2-12　The order of a reaction is determined by ＿＿.

A）The concentration of the reactants

B）The activation energy

C）The reaction rate constant

D）The temperature

2-13　The activation energy is ＿＿ proportional to the reaction rate.

A）directly

B）inversely

C）not

D）cannot be determined from the information given

2-14　The Arrhenius equation relates the rate constant to ＿＿.

A）the order of the reaction

B）the activation energy and the temperature

C）the concentration of the reactants

D）the reaction rate

2-15　Complex reaction kinetics equations may involve all of the following except ＿＿.

A）reversible reactions

B）parallel reactions

C）sequential reactions

D）homogenous and heterogeneous reactions

2.3 Fundamentals of Reactor Design and Basic Equations 反应器设计基本内容和基本方程

Reactors are vessels that facilitate chemical reactions by providing a suitable environment for reactants to interact. Reactor design involves selecting the appropriate reactor type（反应器类型），determining the optimal operating conditions（理想操作条件），and calculating the required reactor volume（反应器体积）.

To select the proper reactor type，various factors such as reaction mechanism（反应机理），reaction kinetics，and the properties of the reactants and products need to be considered. Once the reactor type is selected，optimal operating conditions such as temperature，pressure，and reactant concentration need to be determined for maximum yield and selectivity.

Calculating the required reactor volume is essential in reactor design to ensure that the reaction is economically viable. The reactor volume is dependent on the reaction rate，the desired conversion rate，and the residence time of the reactants in the reactor.

The basic equations（基本方程）involved in reactor design include material balance equations（物料平衡方程），heat balance equations（热量平衡方程），momentum balance equations（动量平衡方程），and kinetic rate equations（动力学速率方程）. Material balance equations describe the concentration changes of reactants and products in the reactor，while heat balance equations describe the heat transfer involved in the reactor. Momentum balance equations describe the pressure changes in the reactor，and kinetic rate equations describe the changes in the reaction rate.

In conclusion，reactor design is an important aspect of chemical engineering and involves selecting the appropriate reactor type，determining optimal operating conditions，and calculating the required reactor volume. Understanding the basic equations involved in reactor design is crucial in designing efficient and cost-effective reactors.

技术理论

反应器设计的基本方程包括：物料衡算方程、热量衡算方程、动量衡算方程以及上节提到的化学反应动力学方程。在计算恒温反应器的反应速率时，可以不考虑热量平衡方程，而动量平衡方程一般情况下也可以不考虑。

如式（2-10）、式（2-11）所示，分别为物料衡算方程和热量衡算方程。

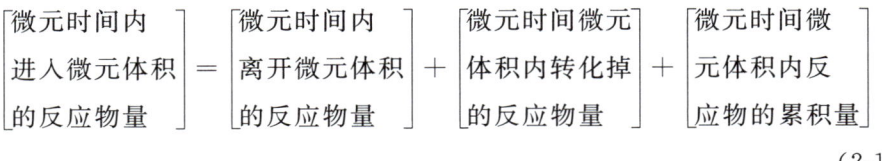

$$(2\text{-}10)$$

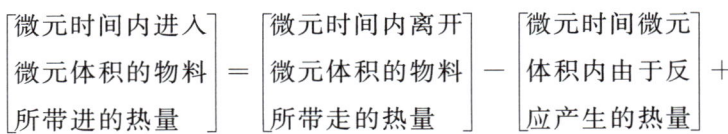

$$(2\text{-}11)$$

关键词详解

物料平衡方程，material balance equation，对进入反应器的反应物进行物料的衡算。

热量平衡方程，heat balance equation，对进入反应器的反应物进行热量的衡算。

动量平衡方程，momentum balance equation，计算进入反应器的压力变化。一般情况下，反应器计算可以不考虑此项反应物进行动量的衡算。

互动练习

2-16　What is the primary goal of reactor design?

A）To facilitate chemical reactions

B）To select the optimal reactor type

C）To determine the operating conditions

D）To calculate the required reactor volume

2-17　What factors need to be considered when selecting a reactor type?

A）Reaction mechanism

B）Reactor properties

C）Reaction kinetics

D）All of the above

2-18　What are the optimal operating conditions that need to be determined in reactor design?

A）Reactor volume

B）Reaction rate

C）Residence time

D）Temperature，pressure，and reactant concentration

2-19　Which of the following equations is NOT involved in reactor design?

A）Material balance equations

B）Energy balance equations

C）Momentum balance equations

D）Kinetic rate equations

2-20　What is the primary purpose of calculating the required reactor volume in reactor design?

A）To determine the optimal reactor type

B）To ensure that the reaction is economically viable

C）To determine the operating conditions

D）To select the appropriate reaction mechanism

2.4　Volume and Number Calculations for BRs
间歇釜式反应器体积和数量计算

BRs are commonly used in the chemical industry for carrying out reactions in batches（分批）. When designing BRs for a given processing volume，it is important to consider factors such as the reactor volume，number of reactors required，and the number of batches（批数）to be operated per day.

If the total processing volume（总处理量），reactor volume，number of batches per day and loading coefficient（装料系数）are given，the number of reactors required can be calculated by dividing the total processing volume by the reactor volume，the loading coefficient，and the number of batches per day. Since the calculated number of reactors is often not an integer，it needs to be rounded up to the nearest whole number to ensure sufficient capacity. The spare factor（后备系数）is the ratio of the rounded-up number of reactors to the calculated number of reactors. For example，if the total processing volume is 10000 liters per day，the reactor volume is 1000 liters，the loading coefficient is 0. 8，and there are 4 batches per day，then the number of reactors required is 3. 125，which would be rounded up to 4 reactors，and the spare factor is then calculated as 4/3. 125＝1. 28.

On the other hand，if the total processing volume，number of batches per day，and number of reactors are given，the reactor volume required can be calculated by dividing the total processing volume by the product of the number of reactors，the loading coefficient，and the number of batches per day. For example，if the total processing volume is 10000 liters per day，there are 10 reactors，the loading coefficient is 0. 8，and there are 4 batches per day，then the reactor volume required for each reactor is 312. 5 liters.

In BR design，it is also important to consider factors such as the reaction kinetics，reactant properties，and the desired conversion rate. Proper consideration of these factors，along with the reactor volume and number of reactors required，can help ensure efficient and effective BRs operation.

技术理论

反应器设计以及生产过程中往往需要计算反应器实际体积。其通常可以通过物料衡算，求出原料处理量，再进一步进行反应釜的体积、装料系数、操作批次、后备系数、操作周期和反应釜的数量的计算。

关键词详解

反应釜的体积，volume of the reactor，分为反应器有效体积 V_R（effective volume）和反应器实际体积 V（actual volume）。其中反应器有效体积按式（2-12）和式（2-13）计算：

$$V_R = \varphi V = V_0(\tau + \tau')$$ （2-12）

$$V = \frac{V_0(\tau + \tau')\delta}{\varphi n}$$ （2-13）

式中，V 为反应器实际体积，V_0 为单位时间体积处理量，τ 为反应时间，τ' 为辅助时间，δ 为后备系数，φ 为装料系数，n 为反应釜的圆整数目。

装料系数，loading coefficient，用 φ 表示，等于 V_R/V。常见设备的装料系数如表 2-3 所示。

表 2-3　常用设备装料系数

条件	装料系数φ 范围
不带搅拌或搅拌缓慢的反应釜	0.8～0.85
带搅拌的反应釜	0.7～0.8
易起泡沫和在沸腾下操作的设备	0.4～0.6
贮槽和计量槽（液面平静）	0.85～0.9

操作周期，operating cycletime，用 t 表示，为指生产每一批物料的全部操作时间：

$$t = \tau + \tau' \qquad (2\text{-}14)$$

式中，τ 为反应时间；τ' 为辅助时间（装料、卸料、检查及清洗设备等所需时间）。

操作批次，operation batch，包括每天操作批次 α 和每只釜每天操作批次 β：

$$\alpha = \frac{24V_0}{V_R} = \frac{24V_0}{V\varphi} \qquad (2\text{-}15)$$

$$\beta = \frac{24}{t} = \frac{24}{\tau + \tau'} \qquad (2\text{-}16)$$

反应釜的数量，number of reactors，用 n' 表示：

$$n' = \frac{\alpha}{\beta} = \frac{V_0(\tau + \tau')}{\varphi V} \qquad (2\text{-}17)$$

后备系数，spare factor，由式（2-17）计算得到的 n' 值通常不是整数，需圆整成整数 n。这样反应釜的生产能力较计算要求提高了，其提高程度称为生产能力的后备系数（一般在 1.1～1.15 较为合适），以 δ 表示：

$$\delta = \frac{n}{n'} \qquad (2\text{-}18)$$

互动练习

2-21　What is the loading coefficient in BR design?

A）The number of reactors required

B）The number of batches per day

C）The ratio of the processing volume to the reactor volume

D）The ratio of the number of reactors to the processing volume

2-22　How to calculate the number of reactors required?

A）By dividing the reactor volume by the processing volume

B）By dividing the processing volume by the loading coefficient，the reactor volume，and the number of batches per day

C）By multiplying the reactor volume by the number of batches per day and the loading coefficient

D）By subtracting the loading coefficient from the number of batches per day and dividing the result by the processing volume

2-23　What is the formula for calculating the reactor volume required for each reactor?

A）Total processing volume/reactor volume

B）Total processing volume/（number of reactors × loading coefficient × number of batches per day）

C）Number of reactors × loading coefficient × number of batches per day/total processing volume

D）Reactor volume × number of batches per day × loading coefficient/total processing volume

2-24　What factors should be considered in BR design?

A）Only the reactor volume and number of reactors

B）Reaction kinetics，reactant properties，and desired conversion rate

C）Only the processing volume and loading coefficient

D）The number of batches per day and the desired reaction time

2-25　How can efficient and effective BRs operation be ensured?

A）By using many reactors

B）By operating at maximum reactor volume

C）By considering reaction kinetics，reactant properties，and desired conversion rate

D）By increasing the number of batches per day

2.5 Kinetic Calculation Method for BRs
间歇釜式反应器动力学计算方法

For constant volume and constant temperature BRs，the BR design equation can be used to calculate the reaction time（反应时间）required to achieve a desired conversion. This equation involves the initial reactant concentrations，the rate constant of the reaction，and the desired conversion. The equation can also be used to determine the maximum conversion achievable within a given reaction time.

In the case of constant volume and non-isothermal（非恒温）BRs，the energy balance equation is added to the batch reactor design equation. This equation takes into account the heat of reaction and the heat transfer between the reactor and the surroundings. The resulting equation can then be solved to obtain the reaction time and the maximum conversion achievable under non-isothermal conditions.

技术理论

间歇反应是非定态操作，釜内组分的浓度随反应时间而变化，如图 2-4 所示。

间歇釜式反应器内物料状态均一，且反应期间没有物料进出，故根据式（2-10）有：

$$(-r_A)V_R d\tau + dn_A = 0 \qquad (2-19)$$

即

$$\tau = n_{A0} \int_{x_{A0}}^{x_{Af}} \frac{dx_A}{(-r_A)V_R} \qquad (2-20)$$

式（2-20）在恒容、恒温、非恒温情况下其结果均可做简化。将动力学方程（常见一级不可逆反应或者二级不可逆反应）代入，可求得相应计算反应器体积和转化率的关系。

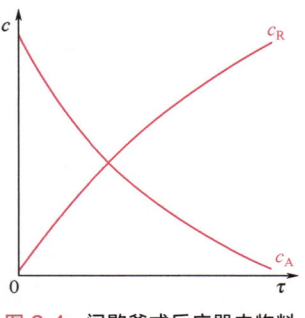

图 2-4　间歇釜式反应器内物料浓度随时间变化关系

关键词详解

恒容，constant-volume，多数液相反应都认定为恒容，此时式（2-20）可简化为：

$$\tau = c_{A0} \int_{x_{A0}}^{x_{Af}} \frac{dx_A}{(-r_A)} \qquad (2-21)$$

$$\tau = -\int_{c_{A0}}^{c_A} \frac{dc_A}{(-r_A)} \qquad (2-22)$$

由式（2-21）可知，间歇操作釜式反应器达到一定转化率所需的反应时间，只取决于过程的反应速率，而与反应器的大小无关。

恒温，isothermal，恒温情况下的反应速率常数为定值，其动力学方程式代入式（2-20）或式（2-21），便可求得反应时间和转化率的关系。当动力学方程解析式相当复杂或不能做数值积分时，可以图解积分法计算所需反应时间，如图 2-5 所示。

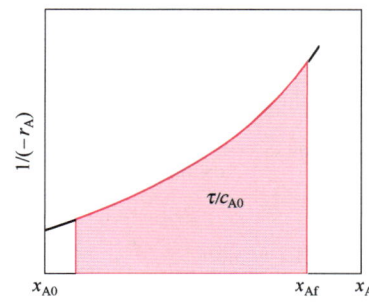 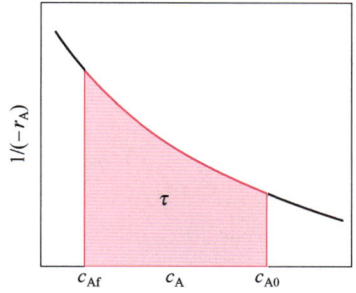

图 2-5　间歇釜式反应器恒温过程图解计算

非恒温，non-isothermal，间歇反应过程温度与时间的关系可由热量衡算式来确定。反应期间没有物料进出；当反应在绝热条件下进行时，传热项为零。故根据式（2-11）有：

$$(T - T_0) = \lambda(x_{Af} - x_{A0}) \tag{2-23}$$

式中，$\lambda = \dfrac{(-\Delta H_A)\, n_{A0}}{m_t c_{pt}}$，称绝热温升。

把式（2-23）代入式（2-22），则式（2-22）变成只含有 x_A 的微分方程，解此微分方程即可得到反应时间，或用图解法求得反应时间，如图 2-6 所示。

绝热温升，adiabatic temperature rise，其意义为当反应系统中的组分 A 全部转化时，系统温度变化的度数。

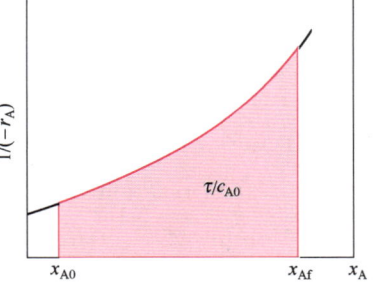

图 2-6　间歇釜式反应器非恒温
过程图解计算

一级不可逆反应，reaction of irreversible one-order，动力学方程式为 $(-r_A) = kc_A$，在恒温条件下 k 为常数，而恒容条件下有 $c_A = c_{A0}(1 - x_A)$，并将其代入式（2-20）得：

$$V_R = V_0\tau = c_{A0}V_0 \int_{x_{A0}}^{x_{Af}} \frac{dx_A}{kc_{A0}(1 - x_A)} \tag{2-24}$$

$$= \frac{V_0}{k} \ln \frac{1 - x_{A0}}{1 - x_{Af}}$$

二级不可逆反应，reaction of irreversible two-order，动力学方程式为 $(-r_A) = kc_A^2$，若 $x_{A0} = 0$，同理可得：

$$V_R = V_0\tau = c_{A0}V_0 \int_0^{x_{Af}} \frac{dx_A}{kc_{A0}^2(1 - x_A)^2} \tag{2-25}$$

$$= V_0 \frac{x_{Af}}{kc_{A0}(1 - x_{Af})}$$

互动练习

2-26　What is the main factor that determines the reaction time needed to achieve a desired conversion in a constant-volume，constant-temperature BR?

A）Reactor volume

B）Initial reactant concentrations

C）Heat of reaction

D）Temperature of the surroundings

2-27　In the case of non-isothermal BRs，what additional factor must be considered in the design equation?

A）Pressure change

B）Heat transfer between the reactor and the surroundings

C）Change in concentration of reactants

D）Catalyst efficiency

2-28　In a constant-volume，isothermal BR，the reaction time required to reach a certain conversion only depends on：

A）Reactor size

B）Reaction rate

C）Heat of reaction

D）Volume of reactants

2-29　What is the primary purpose of the adiabatic temperature rise in a non-isothermal BR?

A）To increase the pressure within the reactor

B）To measure the heat loss during the reaction

C）To determine the temperature change when all reactants are converted

D）To adjust the rate constant

2-30　Which of the following is true for a first-order irreversible reaction in a constant-volume BR?

A）The rate constant is constant under non-isothermal conditions

B）The reaction rate is dependent on the volume of the reactor

C）The reaction time is calculated using the equation for a first-order irreversible reaction

D）The reaction time is independent of the concentration of reactants

2.6 Calculation of Diameter and Height for BRs
间歇釜式反应器直径和高度的计算

In order to design an efficient and cost-effective（性价比高的）BR，it is necessary to calculate its diameter（直径）and height. One of the key considerations in determining the dimensions of a stirred BR is the ratio of its height to diameter，which is typically around 1. 2. This ratio ensures optimal mixing and mass transfer，which are crucial for efficient reaction kinetics.

Another important factor to consider when calculating the dimensions of a BR is the volume of the reactor's end caps（反应釜盖）or heads. This volume can be approximated using the equation $V = 0.131D^3$，where V is the volume of the end cap in cubic meters and D is the diameter of the reactor in meters. This equation is particularly useful when designing a reactor with hemispherical end caps（半球形反应釜盖），which are commonly used in chemical industry.

Once the dimensions of the reactor and the end caps are calculated，other factors such as the volume of the agitator blades（搅拌器叶片）and the required head space（顶部空间）above the liquid level，should also be considered. By taking all of these factors into account，it is possible to design an efficient and effective BRs that can meet the desired process requirements while minimizing cost and maximizing productivity.

技术理论

反应釜体积由反应釜圆筒体积和封头体积二者组成。依据反应能力要求，通过工艺计算可以得到反应釜体积；再根据反应釜体积去求得反应釜的圆筒高度与直径（一般搅拌反应釜的高度与直径之比 $H/D = 1.2$ 左右，如图2-7所示）。圆筒高度及直径需要圆整，并检验装料系数是否合适。

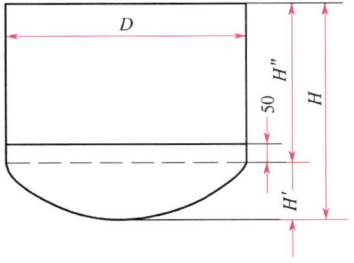

图 2-7　反应釜的主要尺寸

确定了的反应釜的主要尺寸后，其壁厚、法兰尺寸以及手孔、视镜、工艺接管口等均可按工艺条件由国家或行业标准中选择。

关键词详解

反应釜体积，reactor volume，釜盖与釜底体积之和，可按式（2-26）求得。一般在工艺计算决定了反应器的体积后，按该式求得反应釜直径与高度。

封头体积，head volume，$V=0.131D^3$，如图 2-8 所示。

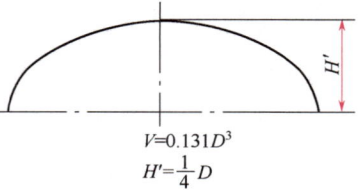

$V=0.131D^3$
$H'=\frac{1}{4}D$

图 2-8 椭圆形封头

$$V=\frac{\pi}{4}D^2H''+0.131D^3 \qquad (2-26)$$

互动练习

2-31 What is the typical ratio of height to diameter in a stirred BR?

A）0.5

B）1.0

C）1.2

D）2.0

2-32 Why is the ratio of height to diameter important in a BR design?

A）It determines the shape of the end caps

B）It ensures optimal mixing and mass transfer

C）It affects the volume of the agitator blades

D）It determines the required head space above the liquid level

2-33 What is the equation used to approximate the volume of the end caps in BRs with hemispherical end caps?

A）$V=0.131D^2$

B）$V=0.131D^3$

C）$V=0.131D$

D）$V=0.131D^4$

2-34 What other factors should be considered when calculating the dimensions of a BR?

A）The volume of the agitator blades and the required head space

B）The chemical composition of the reactants

C）The temperature and pressure conditions

D）The viscosity and density of the reactants

2-35　Why is it important to design an efficient and effective BR?

A）To maximize the cost of production

B）To minimize the productivity of the process

C）To meet the desired process requirements

D）To maximize the size of the reactor

2.7 Balances between Chemical Production Equipment 化工生产设备之间的平衡

Chemical production involves a sequential of unit operations that are inter-connected and need to be balanced to achieve optimal performance. The balance between different pieces of equipment is critical in ensuring that the process is operating efficiently and safely.

Firstly，the balance between reaction vessels （反应釜之间的平衡）is important. For example，if one reactor is producing more than another，it can lead to a buildup （积累）of reactants and products，which can result in the slowing down or stopping of the reaction. The flow rate and temperature must be adjusted to achieve the right balance between reactors.

Secondly，the balance between reaction vessels and physical process equipment （反应釜与物理过程设备之间的平衡）is essential. This can include heat exchangers，distillation columns，or filters. The flow rate and temperature must be optimized to ensure that the reaction and physical process equipment operate in tandem （一前一后地）. If one piece of equipment is working too fast or too slow，it can negatively impact the overall process.

Finally，the balance between reaction vessels and storage tank or metering tank （反应釜与贮槽之间的平衡）must also be considered. If the storage tank is too small or fills up too quickly，it can result in a buildup of reactants and products，which can lead to safety hazards. Additionally，the metering tank must be accurately calibrated to ensure that the correct number of reactants and products are being added or removed from the reaction vessel.

In summary，balancing chemical production equipment is critical in achieving efficient and safe operation. Factors such as flow rate，temperature，and storage capacity must be carefully considered and adjusted to achieve the best results.

技术理论

在通常情况下，加料、出料、清洗等辅助时间是不会太长的。但当前后工序设备之间出现不平衡时，就会大大延长辅助操作的时间。关于设备之间的平衡，大致有下列几种情况：

（1）反应釜与反应釜之间的平衡；

（2）反应釜与物理过程设备之间的平衡；

（3）反应釜与贮槽之间的平衡；

（4）反应釜与计量槽之间的平衡。

关键词详解

反应釜之间的平衡，the balance between reactors，为了便于生产的组织管理和产品的质量检验，通常要求不同批次的物料不相混，这样就应使各道工序每天操作的批次 α 相同。计算时一般首先确定主要反应工序的每天操作批次，然后使其他工序的每天操作批次都与其相同，再确定各工序的设备体积与数量。

反应釜与物理过程设备之间的平衡，the balance between reactor and physical process equipment，当反应后需要过滤或离心脱水时，通常每只反应釜配置一台过滤机或离心机。若过滤需要的时间很短，也可以两只或几只反应釜合用一台过滤机。若过滤需要时间较长，则可以按反应工序的 α 值取其整数倍来确定过滤机的台数，也可以每只反应釜配两台或更多的过滤机（此时可考虑采用一个较大规格的过滤机）。

反应釜与贮槽之间的平衡，the balance between the reactor and the storage tank，当反应后需要浓缩或蒸馏时，因为它们的操作时间较长，通常需要设置中间贮槽，将反应完成液先贮入贮槽中，以避免两个工序之间因操作上不协调而耽误时间。

反应釜与计量槽之间的平衡，the balance between the reactor and the metering tank，通常液体原料都要经过计量后加入反应釜，每只反应釜单独配置专用的计量槽，操作方便。计量槽的体积通常按一批操作需要的原料用量来决定。贮槽的体积则可按一天的需用量来决定。当每天的用量较少时，也可按 $2\sim3$ 天的量来计算。

互动练习

2-36　What is the term used to describe the balance between two reaction vessels?

A）Vessel balance

B）Reaction balance

C）Inter-vessel balance

D）Reactor balance

2-37　What is the term used to describe the balance between a reactor and a physical process equipment?

A）Chemical balance

B）Process balance

C）Physical balance

D）Reactor-process balance

2-38　Which of the following is NOT an example of reactor balance?

A）Balance between two reactors

B）Balance between a reactor and a physical process equipment

C）Balance between a reactor and a storage tank

D）Balance between a reactor and a metering tank

2-39　Why is it important to maintain balance between different chemical production equipment?

A）To ensure the safety of operators

B）To improve the efficiency of the production process

C）To prevent equipment failure

D）All of the above

2-40　Which of the following statements is true regarding reactor balance?

A）It is only necessary for continuous reactors

B）It ensures that the reactants are properly mixed and reacted

C）It is not important for batch reactors

D）It is only important for small-scale production

任务3

Design of CSTRs连续釜式反应器设计

任务要点

本任务重点介绍连续釜式反应器的设计与优化，包括物料平衡方程的应用以及单釜与多釜串联设计的特点。读者将掌握如何通过调整工艺参数优化反应器性能，从而提升连续生产工艺的效率，推动节能减排。

学习目标

知识目标

（1）熟悉连续釜式反应器的结构和运行原理。

（2）掌握连续搅拌釜式反应器的物料平衡计算方法。

技能目标

（1）能进行单个及多釜串联连续釜式反应器的设计。

（2）能分析和优化反应器的操作条件。

价值目标

（1）提升连续生产工艺的效率。

（2）强调可持续发展的工艺设计理念。

3.1　Design of a Single Continuous Stirred Tank Reactor 单个连续釜式反应器设计

In the design of a single continuous stirred tank reactor（CSTR）（连续釜式反应器），it is important to consider the material balance for a particular component. The material balance equation can be expressed as the input rate（进料速率）of a component equals the sum of the output rate（出料速率），reaction rate（反应速率），and accumulation rate（累计速率）. In other words, the rate of material entering the reactor must be equal to the rate of material leaving the reactor plus the rate of material consumed by the reaction and the rate of material accumulated in the reactor.

This equation is important for determining the appropriate reactor size and operating conditions for a given reaction. By analyzing the material balance equation for a specific component，engineers can calculate the necessary reactor volume，flow rates，and other parameters to optimize reactor performance.

Additionally，the material balance equation can be used to monitor and control the reaction process in real-time（实时地）. By measuring the input and output rates of a particular component，engineers can adjust reactor conditions to ensure that the reaction proceeds as intended and that product quality is maintained.

Overall，the material balance equation is a critical tool for designing and optimizing CSTRs in chemical engineering applications.

技术理论

连续釜式反应器的结构和间歇釜式反应器相同，但进出物料的操作是连续的，如图3-1所示。这样的流动状况很接近理想混合流动模型。

在连续釜式反应器内，过程参数与空间位置、时间无关，各处的物料组成和温度都是相同的，且等于出口处的组成和温度。图3-2为单个连续釜式反应器的性能示意图。

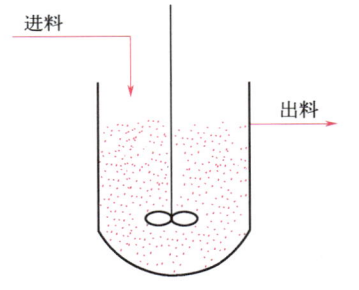

图 3-1　理想混合连续釜式反应器示意图

动画

CSTR进出料

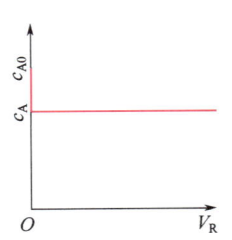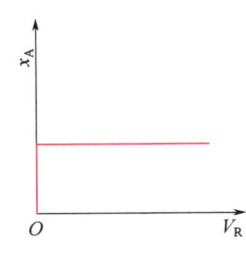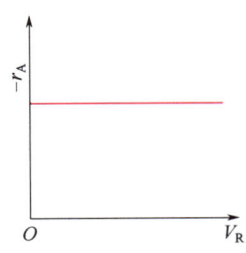

图 3-2　理想混合反应釜的性能

连续反应釜内物料状态均一，且反应期间没有物料积累以及反应参数变化，故根据式（2-10）有：

$$F_{A0} = F_A + (-r_A)V_R$$

即：

$$\frac{V_R}{F_{A0}} = \frac{\bar{\tau}}{c_{A0}} = \frac{\Delta x_A}{(-r_A)} = \frac{x_{Af} - x_{A0}}{(-r_A)} = \frac{x_{Af}}{(-r_A)} \tag{3-1}$$

或

$$\bar{\tau} = \frac{V_R}{V_0} = \frac{c_{A0} - c_A}{(-r_A)} = \frac{c_{A0} x_{Af}}{(-r_A)} \tag{3-2}$$

式中，F_{A0} 为进口物料中组分 A 的摩尔流量，kmol/h；F_A 为出口物料中组分 A 的摩尔流量，kmol/h；V_0 为进口物料体积流量，m^3/h；$\bar{\tau}$ 为物料在釜式反应器中的平均停留时间，h。

关键词详解

摩尔流量，molar flow rate，用 F 表示，单位时间内流入/流出反应器的物料摩尔量。

体积流量，volume flow rate，用 v 表示，单位时间内流入/流出反应器的物料体积，摩尔流量除以体积流量等于浓度。

平均停留时间，average residence time，用 $\bar{\tau}$ 表示，反应物料停留在反应器内的平均时间，等于反应器有效体积除以体积流量。

互动练习

3-1　What is the purpose of the material balance calculation for a continuous stirred tank reactor（CSTR）?

A）To determine the overall reaction rate

B）To calculate the required reactor size

C）To determine the residence time of reactants in the reactor

D）To calculate the amount of reactants needed for the reaction

3-2　What is the equation for material balance in a CSTR for a single component?

A）In＝Out＋R_{xn}（反应量）－Accumulation

B）In－Out＋R_{xn}＋Accumulation＝0

C）In＋Out－R_{xn}＋Accumulation＝0

D）In－Out－R_{xn}＋Accumulation＝0

3-3　What does the term "residence time" refer to in the context of a CSTR?

A）The time it takes for reactants to enter and exit the reactor

B）The time it takes for the reactor to reach steady-state conditions

C）The time it takes for reactants to react completely in the reactor

D）The time it takes for the reactor to be cleaned between batches

3-4　How is the required reactor size calculated for a CSTR?

A）By multiplying the residence time by the flow rate

B）By using the material balance equation to determine the required volume

C）By calculating the reaction rate constant and the stoichiometry of the reaction

D）By using the ideal gas law to determine the required volume

3-5　What is the purpose of the material balance calculation in a CSTR design?

A）To determine the required reactor size and the reaction rate

B）To calculate the reaction rate constant and the stoichiometry of the reaction

C）To determine the residence time and the flow rate of reactants

D）To balance the inflow and outflow of reactants and products in the reactor

3.2　Design of Multiple Sequential CSTRs 多个串联连续釜式反应器设计

Multiple sequential CSTR design involves determining the number of reactors required and their respective sizes to achieve the desired reaction conversion. There are two commonly used methods：analytical（解析的）and graphical（作图的）.

The analytical method（解析法）involves using mathematical equations to calculate the required number of reactors and their respective volumes based on the desired conversion rate，initial concentrations（初始浓度），and flow rates（流速）. This method is more complex and time-consuming but provides accurate results.

The graphical method（作图法）involves plotting conversion rate curves for each reactor stage and determining the number of reactors required to achieve the desired conversion rate（目标转化率）. This method is simpler but less accurate than the analytical method.

To determine the number of reactors required using the graphical method，one must know the processing volume，initial concentrations，and desired conversion rate. If the reactors are assumed to have the same volume，the number of reactors can be determined from the intersection of the conversion rate curves with the desired conversion rate line. To determine the effective volume of each reactor，the processing volume is divided by the number of reactors required.

In summary，the design of multiple sequential CSTRs involves the use of analytical or graphical methods to determine the required number of reactors and their respective volumes to achieve the desired conversion rate. The analytical method is more complex but provides more accurate results，while the graphical method is simpler but less accurate.

技术理论

单个连续釜式反应器返混达最大，反应效率偏低，工业上通常采用多釜串联连续操作反应器，来解决这个问题（图 3-3）。

动画

多釜串联
操作

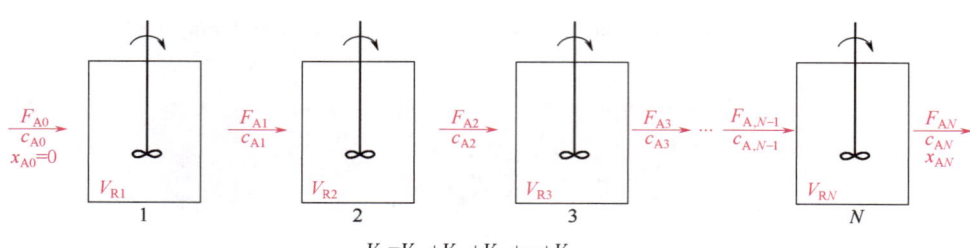

$$V_R = V_{R1} + V_{R2} + V_{R3} + \cdots + V_{RN}$$

图 3-3 多釜串联连续操作反应器操作示意图

多釜串联连续操作反应器的计算一般包括多釜串联解析法和多釜串联作图法。

关键词详解

多釜串联连续操作反应器，multiple sequential CSTRs，该反应器通过横向切割，减少返混影响的同时，又可以发挥连续操作釜反应器自身的优点；另外还可以在各釜内控制不同的反应温度和物料浓度以及不同的搅拌和加料情况，以适应工艺上的不同要求。

多釜串联解析法，multiple-sequential tank analytical method，假设多釜串联连续操作反应器中各釜内均为理想混合，有：

$$F_{A(i-1)} = F_{Ai} + (-r_A)_i V_{Ri} \tag{3-3}$$

即：

$$\frac{V_{Ri}}{V_0} = \frac{c_{A(i-1)} - c_{Ai}}{(-r_A)_i} \tag{3-4}$$

式中，$\dfrac{V_{Ri}}{V_0}$ 为物料在第 i 釜内的平均停留时间，以 $\overline{\tau}_i$ 表示。

反应器的总有效体积为 $V_R = V_{R1} + V_{R2} + \cdots + V_{Ri} + \cdots + V_{RN}$

多釜串联作图法，multiple-sequential tank graphical method，首先根据动力学方程式或实验数据绘出在操作温度下的 $(-r_A) = k c_A^n$ 的动力学关系曲线

（如图 3-4 中的 OA 线）；然后过横坐标上初始浓度 c_{A0} 作一斜率为 $-\dfrac{1}{\tau_i}$ 的直线，与动力学关系曲线所得交点对应的坐标值，即为多釜串联中第一釜的出口浓度 c_{A1}。以该浓度为初始浓度，同样即可得出第二釜的出口浓度。以此类推，即可得出后面各釜的浓度。

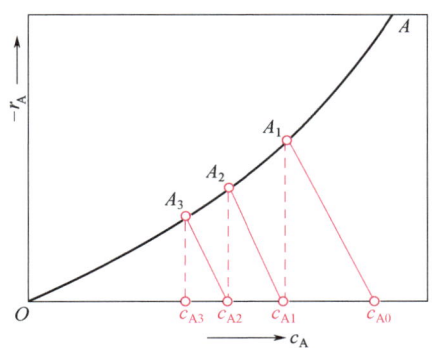

图 3-4　多釜理想连续反应器的图解计算

▶ 动画

多釜理想连续反应器的图解计算

如果串联的各反应釜的有效体积不同，则物料通过各釜的平均停留时间也不同，即各釜操作线斜率 $-\dfrac{V_0}{V_{Ri}}$ 不同，此时就需要以各釜的操作线与对应的动力学曲线相交，计算各釜的出口浓度和串联的台数。

互动练习

3-6　What is the purpose of using multiple sequential CSTRs in chemical reactions?

A）To increase the reaction time

B）To increase the efficiency of the reaction

C）To reduce the reaction temperature

D）To reduce the reaction pressure

3-7　Which method can be used to determine the number of CSTRs needed for a given processing volume，initial concentration，and desired final conversion rate?

A）Analytical method

B）Graphical method

C）Numerical method

D）Empirical method

3-8　In the graphical method for determining the number of sequential-connected CSTRs，what does the slope of the conversion vs reactor volume plot represent?

A）Reaction rate

B）Conversion rate

C）Residence time

D）Number of reactors

3-9　For a given processing volume，initial concentration，and desired final conversion rate，which of the following factors can affect the number of CSTRs needed?

A）The temperature of the reaction

B）The pressure of the reaction

C）The size of the CSTRs

D）The type of catalyst used

3-10　What is the advantage of using multiple sequential-connected CSTRs compared to a single large CSTR for a given reaction process?

A）Lower cost

B）Higher efficiency

C）Lower energy consumption

D）Smaller footprint

Design and Selection of Supporting Facilities for Tank Reactors
釜式反应器配套设施设计与选择

本任务涉及釜式反应器配套设施（如搅拌器、换热装置及配管系统）的设计与选择。读者将学习如何根据工艺需求设计合适的配套设施，优化其布局和功能，提高反应效率，降低能耗，并保证生产过程的安全性。

学习目标

知识目标

（1）了解釜式反应器配套设施（如搅拌器、换热装置及配管系统）的组成和功能。

（2）掌握配套设施的设计原则及选择标准。

（3）认识不同工艺对配套设施的特殊需求。

技能目标

（1）能根据生产需求设计合适的搅拌系统、换热系统和配管系统。

（2）能优化配套设施以提高反应效率和产品质量。

价值目标

（1）注重设施设计的安全性和经济性。

（2）强调环保设施的重要性，减少生产过程中的资源浪费。

4.1 Design and Selection of Agitation System for Tank Reactors
搅拌装置设计与选择

Mixing（混合）is a critical process in many industries such as chemical，pharmaceutical，and food processing. The purpose of mixing is to achieve homogeneity，uniformity，and desired reaction rates of the mixture. The mixing requirements depend on the nature of the materials being mixed，the desired product，and the process parameters.

Fluid flow characteristics play a crucial role in the design and selection of mixing equipment. The primary modes of fluid flow are radial flow（轴向流），axial flow（径向流），and tangential flow（切线流）. The flow pattern affects the mixing performance，and different mixing equipment designs are suitable for specific applications.

Several types of mixing equipment are commonly used，each with unique structures and characteristics. Propeller mixers（桨式搅拌器）are simple and cost-effective，while turbine mixers（涡轮式搅拌器）are suitable for high-viscosity liquids. Pusher mixers（推进式搅拌器）are ideal for high-speed mixing，and frame and anchor mixers（框式和锚式搅拌器）are effective for viscous materials. Ribbon and screw mixers（螺带式和螺旋式搅拌器）are suitable for mixing powders and solids.

Choosing the right mixing equipment depends on several factors，such as the viscosity of the material，the purpose of mixing，and the required mixing intensity. Different types of mixers are suitable for different applications，and the selection process must consider the mixing speed，power，and efficiency.

Accessories such as baffles（挡板）and guide tubes（导流筒）can be used to enhance the mixing process by reducing vortex formation（漩涡形成）and directing the flow of the material. Proper installation of these accessories can improve mixing efficiency and reduce processing time.

In summary，designing and selecting the appropriate mixing equipment is essential to achieve the desired product quality and production efficiency.

技术理论

搅拌器设置的目的是加强釜式反应器内物料的均匀混合，以强化传质和传热。具体包括均相液体的混合、液液分散、气液分散、固液分散、固体溶解以及强化传热等。

搅拌器的液液、气液、固液分散等搅拌效果，主要取决于搅拌器所带来的主体对流扩散、涡流对流扩散以及分子扩散。从扩散强度来说，主体对流扩散＞涡流对流扩散＞分子扩散；而对混合程度来说，分子扩散＞涡流对流扩散＞

主体对流扩散。搅拌越剧烈，涡流运动就越强烈，湍流程度就越大，分散程度就越高，即漩涡的尺寸就越小。其中主体对流扩散只能把物料破碎分裂成"微团"。涡流对流扩散可以把微团的尺寸降低到漩涡本身的大小。在通常的搅拌条件下，漩涡的最小尺寸为几十微米。分子水平上的完全均匀混合程度只有通过分子扩散才能达到。

根据分散程度，混合过程可以分为宏观混合和微观混合。不同的搅拌过程对宏观混合和微观混合的要求是不同的。对于某些局部浓度过高会大大降低选择性的反应，要求提高微观混合。对于液液分散或固液分散，不存在相间的分子扩散，只能达到宏观混合，并依靠漩涡的湍流运动减小微团的尺寸。而对于均相液体的混合，由于分子扩散速率很快，混合速率受宏观混合控制，应设法提高宏观混合速率。

液体在设备范围内作循环流动的途径称作液体的"流动模型"，简称"流型"。在搅拌设备中起主要作用的是循环流和涡流，不同的搅拌器所产生的循环流的方向和涡流的程度不同，因此搅拌设备内流体的流型可以归纳成三种：轴向流、径向流以及切线流。

在化学工业中常用的搅拌装置是机械搅拌装置，典型的机械搅拌装置如图 4-1 所示。它包括搅拌器及其辅助部件。

搅拌器是实现搅拌操作的主要部件，其主要的组成部分是叶轮，针对不同的物料系统和不同的搅拌目的出现了许多类型叶轮的搅拌器。如图 4-2 所示。各种搅拌桨的性能比较如表 4-1 所示。

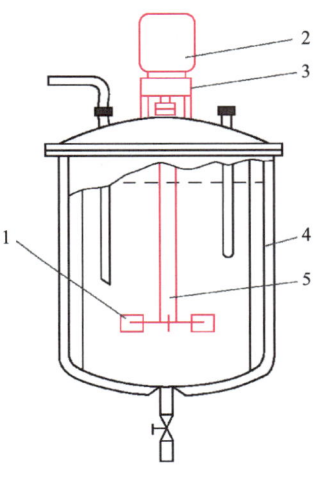

图 4-1 典型的机械搅拌装置

1—搅拌器；2—电机；3—减速箱；
4—挡板；5—搅拌轴

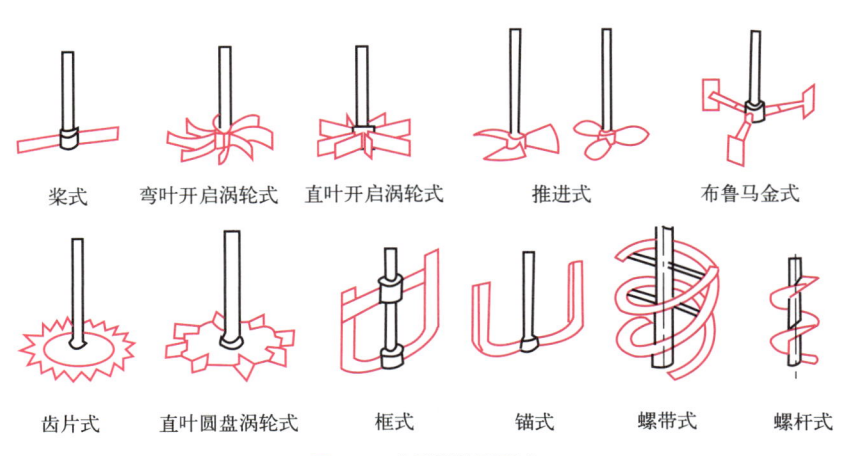

桨式	弯叶开启涡轮式	直叶开启涡轮式	推进式	布鲁马金式

齿片式	直叶圆盘涡轮式	框式	锚式	螺带式	螺杆式

图 4-2 典型搅拌器型式

47

表 4-1　不同叶轮的搅拌器的比较

桨类型	转速	直径	适用场合
桨式搅拌器	较低，一般为 20～80r/min，圆周速度在 1.5～3m/s。	反应釜内径 1/3～2/3，当反应釜直径很大时采用两个或多个桨叶，液体物料层很深时可在轴上装置数排桨叶。	流动性大、黏度小的液体物料，纤维状和结晶状的溶解液。
涡轮搅拌器	速度较大，线速度约为 3～8m/s，转速范围为 300～600r/min。	按照有无圆盘可分为圆盘涡轮搅拌器和开启涡轮搅拌器；按照叶轮又可分为平直叶和弯曲叶两种。	当能量消耗不大时，搅拌效率较高，搅拌产生很强的径向流。它适用于乳浊液、悬浮液等。剪切作用较大。
推进式搅拌器	线速度可达 5～15m/s，转速范围为 300～600r/min。	取反应釜内径的 1/4～1/3。	适用于需要有更大的流速的情况，一般此时反应釜内设有导流筒。
框式（锚式）搅拌器	线速度约 0.5～1.5m/s，转速范围约 50～70r/min。	直径较大，一般取反应器内径的 2/3～9/10	常用于传热、晶析操作和高黏度液体、高浓度淤浆和沉降性淤浆的搅拌。
螺带（螺杆）式搅拌器	通常不超过 50r/min。	常用扁钢按螺旋形绕成，直径较大，常做成几条紧贴釜内壁，与釜壁的间隙很小。	主要用于高黏度液体的搅拌。

搅拌器的选型主要包括根据物料黏度选型、根据搅拌目的选型。

对于低黏度液体，应选用小直径、高转速搅拌器，如推进式、涡轮式；对于高黏度液体，应选用大直径、低转速搅拌器，如锚式、框式和桨式。图 4-3 表明了几种典型的搅拌器随黏度的高低而有不同的使用范围。

动画

黏度对搅拌器选型的影响

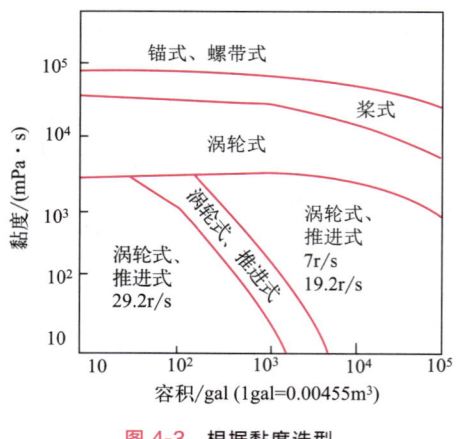

图 4-3　根据黏度选型

根据搅拌目的选型具体选择如表 4-2 所示。

表 4-2　按搅拌目的选型

搅拌目的	特征	控制因素	推荐选用
低黏度均相液体混合	分子扩散速率很快	循环流量	推进式
非均相液液分散	要求增大两相接触表面积、降低两相传质阻力	剪切作用	涡轮式(平直叶)
气液分散过程	要求得到高分散度的"气泡"	剪切作用	涡轮式(平直叶圆盘)
固体悬浮操作	让固体颗粒均匀悬浮于液体	循环流量	推进式(固液密度差小)，开启式涡轮(固液密度差大)
固体溶解	较大的循环流量，较强的剪切作用	剪切作用	开启式涡轮
以传热为主的搅拌操作	循环流量和换热面上的高速流动	剪切作用	涡轮式

搅拌器的选型主要包括根据物料黏度选型、根据搅拌目的选型。

搅拌附件通常指在搅拌罐内为了改善流动状态而增设的零件，如挡板、导流筒等。

关键词详解

搅拌器，stirrer，由搅拌轴和搅拌电机组成，包括旋转的轴和装在轴上的叶轮。

均相液体的混合，mixing of homogeneous liquids，通过搅拌使反应釜中的互溶液体达到分子规模的均匀程度。

液液分散，liquid-liquid dispersion，把不互溶的两种液体混合起来，使其中的一相液体以微小的液滴均匀分散到另一相液体中。被分散的一相为分散相，另一相为连续相。被分散的液滴越小，两相接触面积越大。

气液分散，gas-liquid dispersion，在气液接触过程中，搅拌器把大气泡打碎成微小气泡并使之均匀分散到整个液相中，以增大气液接触面积。另一方面，搅拌还造成液相的剧烈湍动，以降低液膜的传质阻力。

固液分散，solid-liquid dispersion，让固体颗粒悬浮于液体中。例如硝基物的液相加氢还原反应，一般以骨架镍为固体催化剂，反应时需要把固体颗粒催化剂悬浮于液体中，才能使反应顺利进行。

固体溶解，solid dissolution，当反应物之一为固体而溶于液体时，固体颗粒需要悬浮于液体之中。搅拌可加强固液间的传质，以促进固体溶解。

强化传热，heat transfer enhancement，有些物理或化学过程对传热有很高的要求，或需要消除釜内的温度差，或需要提高釜内壁的给热系数，搅拌可以达到上述强化传热的要求。

主体对流扩散，main convection diffusion，是指搅拌器叶轮运转时，其周围液体也跟着以很高的速度运动起来，从而带动所有液体在设备范围内流动。

涡流对流扩散，eddy convection diffusion，是指被叶轮带动的高速旋转的液体对它周围的液体造成强烈的剪切作用，从而产生更多的漩涡，并造成局部范围内的扩散混合作用。

分子扩散，molecular diffusion，是指分子水平上的完全均匀混合。

宏观混合，macro-mixing，在设备范围内呈微团均匀分布的混合过程，主体对流扩散和涡流对流扩散只能进行宏观混合。

微观混合，micro-mixing，达到分子规模分布均匀的混合过程，只有分子扩散才能进行微观混合。加快搅拌可以大大增加分子扩散的表面积，减小分子扩散的距离，提高微观混合速率。

循环流，circulation flow，可以产生液体循环流量，在消耗同等功率的条件下，如果采用低转速、大直径的叶轮，可以增大液体循环流量，同时减少液体受到的剪切作用，有利于宏观混合。

涡流，eddy flow，产生剪切作用，在消耗同等功率的条件下，如果采用高转速、小直径的叶轮，可以增大液体受到的剪切作用，同时减少液体循环流量，有利于微观混合。

轴向流，axial flow，物料沿搅拌轴的方向循环流动，如图4-4（a）所示。凡是叶轮与旋转平面的夹角小于90°的搅拌器转速较快时所产生的流型主要是轴向流。轴向流的循环速度大，有利于宏观混合，适合于均相液体的混合、沉降速度低的固体悬浮。

径向流，radial flow，物料沿着反应釜的半径方向在搅拌器和釜内壁之间的流动，如图4-4（b）所示。径向流的液体剪切作用大，造成的局部涡流运动剧烈，因此，它特别适合需要高剪切作用的搅拌过程，如气液分散、液液分散和固体溶解。

▶ **动画**

搅拌液体的
流型
-轴向
-径向
-切线

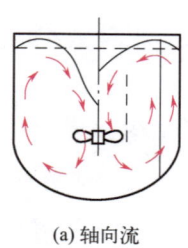

(a) 轴向流

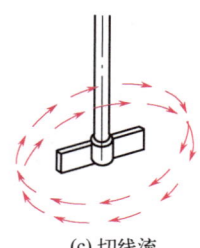

(b) 径向流

(c) 切线流

图 4-4 搅拌液体的流型

切线流，tangent flow，物料围绕搅拌轴作圆周运动，如图 4-4(c) 所示。平桨式搅拌器在转速不大且没有挡板时所产生的主要是切线流。切线流的存在除了可以提高反应釜内壁的对流给热系数外，对其他的搅拌过程是不利的。切线流严重时，液体在离心力的作用下涌向器壁，使器壁周围的液面上升，而中心部分液面下降，形成一个大漩涡，这种现象称为"打漩"，如图 4-5 所示。液体打漩时几乎不产生轴向混合作用，所以一般情况下，应防止打漩。

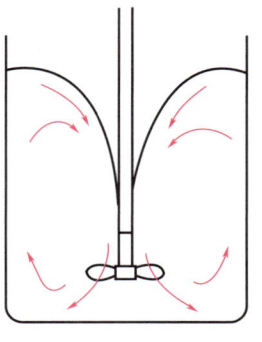

图 4-5　打漩现象

动画

打漩现象

辅助部件，accessories，搅拌器辅助部件包括电机、减速箱、支架、挡板、搅拌轴和导流筒等。

叶轮，impeller，它随旋转轴运动将机械能施加给液体，并促使液体运动。

桨式搅拌器，paddle agitator，由桨叶、键、轴环、竖轴所组成。

涡轮式搅拌器，turbine agitator，涡轮式搅拌器按照有无圆盘可分为圆盘涡轮搅拌器和开启涡轮搅拌器；按照叶轮又可分为直叶和弯叶两种。涡轮搅拌器速度较大，线速度约为 3～8m/s，转速范围为 300～600r/min。涡轮搅拌器的主要优点是当能量消耗不大时，搅拌效率较高，搅拌产生很强的径向流。因此它适用于乳浊液、悬浮液等。剪切作用较大。

推进式搅拌器，propeller agitator，搅拌时能使物料在反应釜内循环流动，液体循环量较大，剪切作用较小。搅拌器的材料常用铸铁和铸钢。推进式搅拌器的标准系列见 HG/T 3796.8—2005。

框式（锚式）搅拌器，frame（anchor）agitator，框式搅拌器可视为桨式搅拌器的变形，即将水平的桨叶与垂直的桨叶连成一体成为刚性的框子，其结构比较坚固，搅动物料量大。

螺带式搅拌器和螺杆式搅拌器，screw belt agitator and screw agitator，螺带式搅拌器，常用扁钢按螺旋形绕成，搅拌时能不断地将粘于釜壁的沉积物刮下来。

根据物料黏度选型，according to the viscosity of materials，在影响搅拌状态的诸物理性质中，液体黏度的影响最大，所以，可根据液体黏度来选型。

根据搅拌目的选型，according to the purpose of mixing，搅拌目的、工艺过程对搅拌的要求是选型的关键。

挡板，baffle，一般是指长条形的竖向固定在罐壁上的板，主要是在湍流状态时为了消除切线流和"打漩"现象而增设的。而在层流状态下，挡板并不影响流体的流动，所以对于低速搅拌高黏度液体的锚式和框式搅拌器来说，安装挡板是毫无意义的。

导流筒，draft tube，导流筒主要用于推进式、螺杆式搅拌器的导流，涡轮式搅拌器有时也用导流筒。导流筒是一个圆筒形，紧包围着叶轮。应用导流筒可使流型得以严格控制，还可得到高速涡流和高倍循环。导流筒可以为液体限定一个流动路线，防止短路；也可迫使流体高速流过加热面以利于传热。对于混合和分散过程，导流筒也能起到强化作用。对于涡轮式搅拌器，导流筒安置在叶轮的上方，使叶轮上方的轴向流得到加强，如图 4-6（a）所示。对于推进式搅拌器，如图 4-6（b）所示，导流筒安置在叶轮的外面，使推进式搅拌器所产生的轴向流得到进一步加强。

导流筒的安装方式

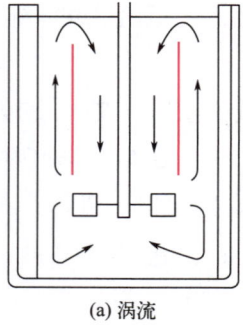　　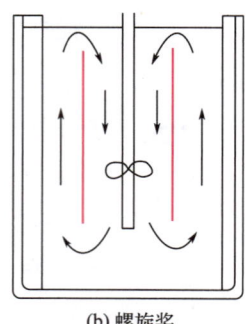

(a) 涡流　　　　　　　　　　　　(b) 螺旋桨

图 4-6　导流筒的安装方式

互动练习

4-1　What is the primary purpose of mixing?

A）To increase the viscosity of the material

B）To achieve homogeneity and uniformity

C）To separate the components of the material

D）To reduce the temperature of the material

4-2　Which of the following is NOT a primary mode of fluid flow in mixing?

A）Axial flow

B）Radial flow

C）Tangential flow

D）Vertical flow

4-3　Which type of mixer is most suitable for high-viscosity liquids?

A）Propeller mixer

B）Turbine mixer

C）Pusher mixer

D）Ribbon mixer

4-4　How should the selection of mixing equipment be based?

A）Material viscosity，mixing purpose，and required mixing intensity

B）Material color，mixing temperature，and mixing speed

C）Material density，mixing pressure，and mixer size

D）Material taste，mixing time，and mixer material

4-5　What is the purpose of using accessories such as baffles and guide tubes in mixing equipment?

A）To increase the temperature of the material

B）To reduce the viscosity of the material

C）To reduce vortex formation and direct the flow of the material

D）To increase the pressure of the material

4.2 Design and Selection of Heat Transfer Equipment for Tank Reactors 釜式反应器的换热装置设计与选择

The design and selection of heat exchangers for tank reactors is an essential aspect of chemical process engineering. Heat exchangers are devices used to transfer thermal energy between two or more fluids at different temperatures. They play a critical role in chemical reactions，especially in tank reactors where heat transfer is crucial for achieving high reaction rates and product yields.

There are several types of heat exchangers commonly used in tank reactors，including types of jacketed （夹套式），coiled （蛇管式），tubular （列管式），external loop （外部循环式），and reflux condensing types （回流冷凝式）. Jacketed heat exchangers are commonly used in small-scale reactions and are suitable for heat transfer in both batch and continuous operations. Coiled heat exchangers are ideal for large-scale reactions and offer excellent heat transfer capabilities due to their extended surface area. Tubular heat exchangers are commonly used in high-pressure and high-temperature applications，while external loop and reflux condenser types are suitable for continuous operations.

The selection of high-temperature heat transfer fluids depends on several factors，including the operating temperature，pressure，and the specific require-ments of the reaction. High-pressure saturated water vapor（高压饱和水蒸气）and high-pressure steam-water mixtures（高压汽-水混合物）are commonly used as heat transfer fluids in tank reactors. Organic heat transfer fluids（有机导热油），molten salts（熔盐），electric heating（电加热），and flue gas（烟道气）are also used in some applications.

The selection of low-temperature heat transfer fluids depends on the cool-ing requirements of the reaction. Cooling water（CW）（冷却水），air，and low-temperature coolants（低温冷却介质）are commonly used as heat transfer flu-ids in tank reactors. Water is the most used coolant in chemical processes due to its high heat transfer efficiency and low cost. Air cooling is often used in ap-plications where water is not readily available，while low-temperature coolants such as liquid nitrogen（液氮）are used in some specialized applications.

In summary，the design and selection of heat exchangers for tank reactors involve several considerations，including the type of heat exchanger，the operat-ing temperature and pressure，and the specific requirements of the reaction.

技术理论

釜式反应器的换热装置包括夹套式换热器、蛇管式换热器、列管式换热器、外部循环式换热器、回流冷凝式换热器等。

常用的加热剂有：高压饱和水蒸气、高压汽-水混合物、有机导热油、熔盐、电加热以及烟道气等。

常用低温冷源有：冷却用水、空气、低温冷却介质等。

关键词详解

夹套式换热器，jacketed heat exchanger，传热夹套一般由钢板焊接而成，它是套在反应器筒体外面能形成密封空间的容器，既简单又方便。夹套内通蒸汽时，其蒸汽压力一般不超过 0.6MPa。当反应器的直径大或者加热蒸汽压力较高时，夹套必须采取加强措施。图 4-7 所示为几种加强的夹套传热结构。

图 4-7 中（a）为一种支撑短管加强的"蜂窝夹套"，可用 1MPa 的饱和水蒸气加热至 180℃。（b）为冲压式蜂窝夹套，可耐更高的压力。（c）和（d）为角钢焊在釜的外壁上的结构，耐压可达到 5～6MPa。

夹套与反应釜内壁的间距视反应釜直径的大小采用不同的数值，一般取

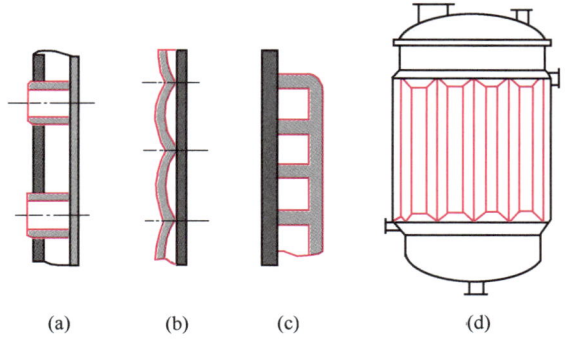

图 4-7　几种加强夹套传热结构

25～100mm。夹套的高度取决于传热面积，而传热面积由工艺要求确定。但须注意夹套高度一般应高于料液的高度，应比釜内液面高出 50～100mm，以保证充分传热。

有时，对于较大型的搅拌釜，为了提高传热效果，在夹套空间装设螺旋导流板如图 4-8 所示，以缩小夹套中流体的流通面积，提高流速并避免短路。螺旋导流板一般焊在釜壁上，与夹套壁有小于 3mm 的间隙。加设螺旋导流板后，夹套侧的传热膜系数一般可由 500W/(m² · K) 增大到 1500～ 2000W/(m² · K)。

图 4-8　螺旋导流板

蛇管式换热器，coiled heat exchanger，当工艺需要的传热面积大，单靠夹套传热不能满足要求时，或者是反应器内壁衬有橡胶、瓷砖等非金属材料时，可采用蛇管、插入套管、插入 D 型管等传热。

工业上常用的蛇管有两种：水平式蛇管如图 4-9 所示和直立式蛇管如图 4-10 所示。排列紧密的水平式蛇管能同时起到导流筒的作用，排列紧密的直立式蛇管同时可以起到挡板的作用，它们对于改善流体的流动状况和搅拌的效果起积极的作用。

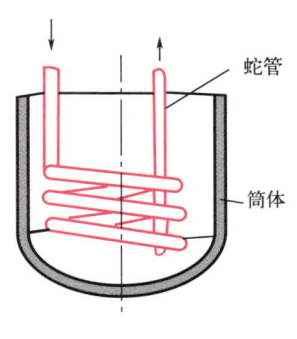

蛇管

筒体

图 4-9　水平式蛇管

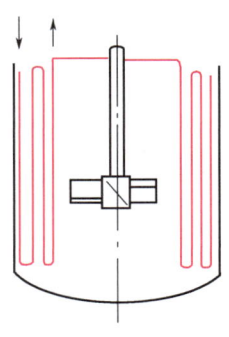

图 4-10　直立式蛇管

蛇管浸没在物料中，热量损失少，且由于蛇管内传热介质流速高，它的给热系数比夹套大得多。但对于含有固体颗粒的物料及黏稠的物料，容易引起物料堆积和挂料，影响传热效果。

工业上常用的几种插入式传热构件如图 4-11 所示。（a）为垂直管，（b）为指型管，（c）为 D 型管。这些插入式结构适用于反应物料容易在传热壁上结垢的场合，检修、除垢都比较方便。

列管式换热器，tubular heat exchanger，对于大型反应釜，需高速传热时，可在釜内安装列管式换热器，如图 4-12 所示。它的主要优点是单位体积所具有的传热面积大，传热效果好，此外结构简单，操作弹性较大。

动画

插入式传热
构件

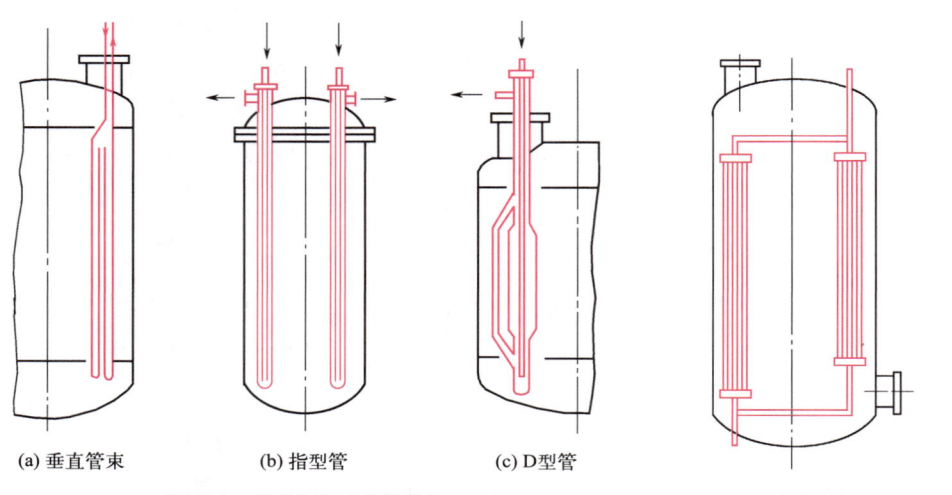

(a) 垂直管束　　　　(b) 指型管　　　　(c) D型管

图 4-11　几种插入式传热构件　　　　图 4-12　内装列管的反应釜

外部循环式换热器，external loop heat exchanger，当反应器的夹套和蛇管传热面积仍不能满足工艺要求，或由于工艺的特殊要求无法在反应器内安装蛇管而夹套的传热面积又不能满足工艺要求时，可以通过泵将反应器内的料液抽出，经过外部换热器换热后再循环回反应器中。

回流冷凝式换热器，reflux condensing heat exchanger，当反应在沸腾温度下进行且反应热效应很大时，可以采用回流冷凝法进行换热，即：使反应器内产生的蒸汽通过外部的冷凝器加以冷凝，冷凝液返回反应器中。采用这种方法进行传热，由于蒸汽在冷凝器中以冷凝的方式散热，可以得到很高的给热系数。

高压饱和水蒸气，high pressure saturated steam，来源于高压蒸汽锅炉、利用反应热的废热锅炉或热电站的蒸汽透平。蒸汽压力可达数兆帕。用高压蒸汽作为热源的缺点是需高压管道输送蒸汽，其建设投资费用大，尤其需远距离输送时热损失也大，很不经济。

高压汽水混合物，high pressure steam water mixture，当车间内有个别设备需高温加热时，设置一套专用的高压汽水混合物作为高温热源，可能是比较经济可行的。这种加热装置的原理如图 4-13 所示。由焊在设备外壁上的高压蛇管（或内部蛇管）1、空气冷却器 2、高温加热炉 3 和安全阀 4 等部分构成一个封闭的循环系统。管内充满 70％的水和 30％的蒸汽，形成汽水混合物。从加热炉到加热设备这一段管道内，蒸汽比例高，水的比例低，而从冷却器返回加

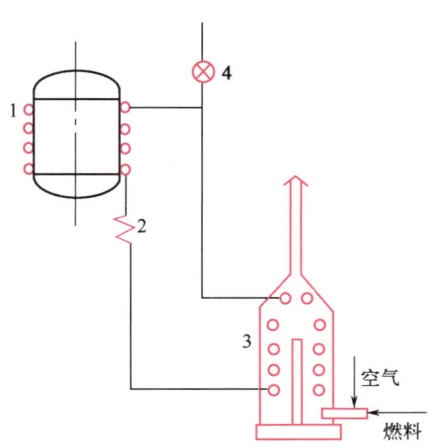

动画

高压汽水混合物的加热装置

图 4-13　高压汽水混合物的加热装置

热炉这一段管道内蒸汽比例低，水的比例高，于是形成一个自然循环系统。循环速度的大小决定于加热的设备与加热炉之间的高位差及汽水比例。

这种高温加热装置适用于 200～250℃的加热要求。加热炉的燃料可用气体燃料或液体燃料，炉温达 800～900℃，炉内加热蛇管用耐温耐压合金钢管。

有机导热油，organic heat transfer fluid，利用某些有机物常压沸点高、熔点低、热稳定性好等特点可提供高温的热源。如联苯道生油、YD、SD 导热油等都是良好的高温载热体。联苯道生油是含联苯 26.5％、二苯醚 73.5％的低共沸点混合物，熔点 12.3℃，沸点 258℃。它的突出优点是能在较低的压力下得到较高的加热温度。在同样的温度下，它的饱和蒸汽压力只有水蒸气压力的几十分之一。

当加热温度在 250℃以下时，可采用液体联苯混合物加热，可有三种加热方案。

① 液体联苯混合物自然循环加热法，如图 4-14 所示。加热设备与加热炉之间保持一定的高位差才能使液体有良好的自然循环。

② 液体联苯混合物强制循环加热法。采用屏蔽泵或者用液下泵使液体强制循环。

③ 夹套内盛联苯混合物，将管状电热器插入液体内的加热法。应用于传热速率要求不太高的场合，如图 4-15 所示。

当加热温度超过 250℃时，可采用联苯混合物的蒸汽加热。根据其冷凝液回流方法的不同，也可分为自然循环与强制循环两种方案。自然循环法设备较简单，不需使用循环泵，但要求加热器与加热炉之间有一定的位差，以保证冷

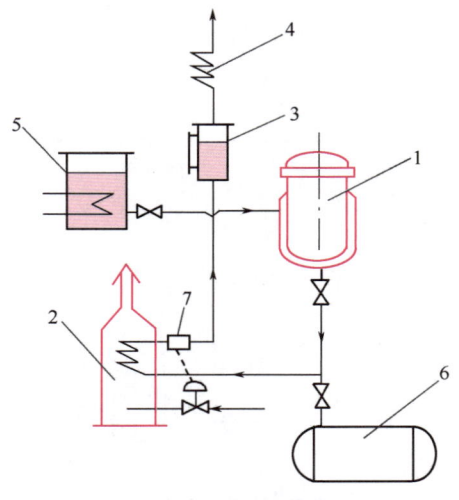

图 4-14 液体联苯混合物自然循环加热装置

1—被加热设备；2—加热炉；3—膨胀器；

4—回流冷凝器；5—熔化炉；6—事故槽；

7—温度自控装置

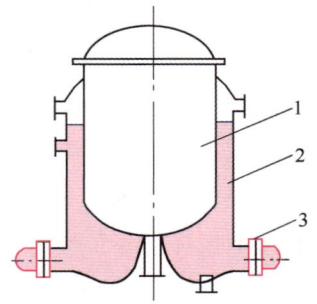

图 4-15 液体联苯混合物夹套浴
电加热装置

1—被加热设备；2—加热夹套；

3—管式电热器

▶ 动画

液体联苯混合物加热装置

−自然循环加热

−夹套浴电加热

−蒸汽夹套浴加热

凝液的自然循环。位差的高低决定于循环系统阻力的大小，一般可取 3～5m。如厂房高度不够，可以适当放大循环液管径以减少阻力。

当受条件限制不能达到自然循环要求时或者加热设备较多、操作中容易产生互相干扰等情况下，可用强制循环流程。

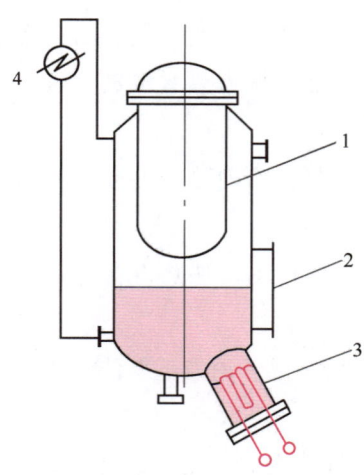

图 4-16 联苯混合物蒸汽夹
套浴加热装置

1—被加热设备；2—液面计；

3—电加热棒；4—回流冷凝器

另一种较为简易的联苯混合物蒸汽加热装置，是将蒸汽发生器直接附设在加热设备上面。用电热棒加热液体联苯混合物，使它沸腾产生蒸汽，如图 4-16 所示。当加热温度小于 280℃，蒸汽压力低于 0.07MPa 时，采用这种方法较为方便。

熔盐，molten salt，反应温度在 300℃ 以上可用熔盐作载热体。熔盐的组成为 $KNO_3 53\%$，$NaNO_3 7\%$，$NaNO_2 40\%$（质量分数，熔点 142℃）。

电加热，electric heating，是一种操作方便、热效率高、便于实现自控和遥控的一种高温加热方法。常用的电加热方法可以分为以下三种类型。

① 电阻加热法。电流透过电阻产生热量实现加热。可采用以下几种结构型式。

a. 辐射加热，即把电阻丝暴露在空气中，借辐射和对流传热直接加热反应釜。此种型式只能适用于不易燃易爆的操作过程。

b. 电阻夹布加热，将电阻丝夹在用玻璃纤维织成的布中，包扎在被加热设备的外壁。这样可以避免电阻丝暴露在大气中，从而减少引起火灾的危险性。但必须注意的是电阻夹布不允许被水浸湿，否则将引起漏电和短路的危险事故。

c. 插入式加热法，将管式或棒状电热器插入被加热的介质中或夹套浴中实现加热（如图 4-15 和图 4-16 所示）。这种方法仅适用于小型设备的加热。

电阻加热可采用可控硅电压调节器自动调节加热温度，实现较为平稳的温度控制。

② 感应电流加热。这是利用交流电路所引起的磁通量变化在被加热体中感应产生的涡流损耗变为热能。感应电流在加热体中透入的深度与设备的形状以及电流的频率有关。在化工生产中应用较方便的是普通的工业交流电产生感应电流加热，称为工频感应电流加热法，它适用壁厚在 5～8mm 以上，圆筒形设备加热（高径比最好在 2～4 以上），加热温度在 500℃ 以下。其优点是施工简便、无明火，在易燃易爆环境中使用比其他加热方式安全、升温快、温度分布均匀。

③ 短路电流加热。将低电压，如 36V 的交流电直接通到被加热的设备上，利用短路电流产生的热量进行高温加热。这种电加热法适用于加热细长的反应器。

烟道气，flue gas，用煤气、天然气、石油加工废气或燃料油等燃烧时产生的高温烟道气作热源加热设备，可用于 300℃ 以上的高温加热。缺点是热效率低、给热系数小、温度不易控制。

冷却用水，cooling water，如河水、井水、城市水厂给水等，水温随地区和季节而变。深井水的水温较低而稳定，一般在 15～20℃。水的冷却效果好，也最为常用。随水的硬度不同，对换热后的水出口温度有一定限制，一般不宜超过 60℃，在不宜清洗的场合不宜超过 50℃，以免水垢的迅速生成。

空气，air，在缺乏水资源的地方可采用空气冷却，其主要缺点是给热系数低，需要的传热面积大。

低温冷却剂，cryogenic coolant，有些化工生产过程需要在较低的温度下进行，这种低温采用一般冷却方法难以达到，必须采用特殊的制冷装置进行人工制冷。

在制冷装置中一般多采用直接冷却方式，即利用制冷剂的蒸发直接冷却冷间内的空气，或直接冷却被冷却物体。制冷剂一般有液氨、液氮等。由于需要额外的机械能量，故成本较高。

有些情况则采用间接冷却方式，即被冷却对象的热量是通过中间介质传送给在蒸发器中蒸发的制冷剂。这种中间介质起着传送和分配冷量的媒介作用，称为载冷剂。常用的载冷剂有三类，即水、盐水及有机物载冷剂。

① 水。比热容大，传热性能良好，价廉易得，但冰点高，仅能用作制取0℃以上冷量的载冷剂。

② 盐水。氯化钠及氯化钙等盐的水溶液，通常称为冷冻盐水。盐水的起始凝固温度随浓度而变，如表4-3所示。氯化钙盐水的共晶温度（−55℃）比氯化钠盐水低，可用于较低温度，故应用较广。氯化钠盐水无毒，传热性能较氯化钙盐水好。

表 4-3　冷冻盐水起始凝固温度与浓度的关系

相对密度（15℃）	氯化钠盐水			氯化钙盐水		
	质量分数/%	100kg 水加盐量/kg	起始凝固温度/℃	质量分数/%	100kg 水加盐量/kg	起始凝固温度/℃
1.05	7.0	7.5	−4.4	5.9	6.3	−3.0
1.10	13.6	15.7	−9.8	11.5	13.0	−7.1
1.15	20.0	25.0	−16.6	16.8	20.2	−12.7
1.175	23.1	30.1	−21.2	—	—	—
1.20	—	—	—	21.9	28.0	−21.2
1.25	—	—	—	26.6	36.2	−34.4
1.286	—	—	—	29.9	42.7	−55.0

氯化钠盐水及氯化钙盐水均对金属材料有腐蚀性，使用时需加缓蚀剂重铬酸钠及氢氧化钠，以使盐水的 pH 值达 7.0～8.5，呈弱碱性。

③ 有机物载冷剂。有机物载冷剂适用于比较低的温度，常用的有如下几种。

a. 乙二醇、丙二醇的水溶液。乙二醇无色无味，可全溶于水，对金属材料无腐蚀性。乙二醇水溶液使用温度可达−35℃（浓度为45%），但用于−10℃（35%）时效果最好。乙二醇黏度大，故传热性能较差，稍具毒性，不宜用于开式系统。

丙二醇是极稳定的化合物，全溶于水，对金属材料无腐蚀性。丙二醇的水溶液无毒；黏度较大，传热性能较差。丙二醇的使用温度通常为−10℃或−10℃以上。

乙二醇和丙二醇溶液的凝固温度随其浓度而变，如表 4-4 所示。

表 4-4　乙二醇和丙二醇溶液的凝固温度与浓度关系

体积浓度/%		20	25	30	35	40	45	50
凝固温度/℃	乙二醇	−8.7	−12.0	−15.9	−20.0	−24.7	−30.0	−35.9
	丙二醇	−7.2	−9.7	−12.8	−16.4	−20.9	−26.1	−32.0

b. 甲醇、乙醇的水溶液

在有机物载冷剂中，甲醇是最便宜的，而且对金属材料不腐蚀，甲醇水溶液的使用温度范围是 0～−35℃，相应的浓度是 15%～40%，在 −20～−35℃ 范围内具有较好的传热性能。甲醇用作载冷剂的缺点是有毒和可以燃烧，在运送、储存和使用中应注意安全问题。

乙醇无毒，对金属不腐蚀，其水溶液常用于啤酒厂、化工厂及食品厂。乙醇也可燃，比甲醇贵，传热性能比甲醇差。

互动练习

4-6　Which type of heat exchanger is commonly used in small-scale reactions?

A）Coiled heat exchanger

B）Jacketed heat exchanger

C）Tubular heat exchanger

D）External loop heat exchanger

4-7　Which of the following is NOT a high-temperature heat transfer fluid commonly used in tank reactors?

A）High-pressure saturated water vapor

B）High-pressure steam-water mixtures

C）Organic heat transfer fluids

D）Cooling water

4-8　Which type of heat exchanger is commonly used in high-pressure and high-temperature applications?

A）Jacketed heat exchanger

B）Coiled heat exchanger

C）Tubular heat exchanger

D）External loop heat exchanger

4-9　Which of the following is NOT a low-temperature heat transfer fluid commonly used in tank reactors?

A）Cooling water

B）Air

C）Liquid nitrogen

D）High-pressure steam-water mixtures

4-10　What is the role of heat exchangers in chemical reactions?

A）To transfer thermal energy between two or more fluids at different temperatures

B）To increase the pressure of the reactants

C）To provide a source of electrical energy

D）To accelerate the reaction rate

任务5

Design of PFRs连续管式反应器设计

任务要点

本任务重点讲解连续管式反应器的设计原理，涵盖物料平衡与热量平衡的基础设计方程式。读者将学习如何通过科学计算优化反应器的尺寸、形状及操作条件，以满足不同化工生产的需求。

学习目标

知识目标

（1）熟悉管式反应器的基本运行原理及分类。

（2）掌握管式反应器设计的基础方程式，如物料平衡方程和热量平衡方程。

（3）了解管式反应器在化工生产中的应用场景及优劣势。

技能目标

（1）能利用设计方程式计算管式反应器的关键参数，如体积、长度和直径。

（2）能分析管式反应器的操作条件并进行优化设计。

价值目标

（1）提高管式反应器设计的效率与科学性。

（2）注重管式反应器设计中的节能减排。

5.1 Basic Design Equations of PFRs 连续管式反应器基础设计方程式

PFRs design involves solving a set of mass and energy balance equations for each component in the reactor.

The mass balance equation for a particular component can be written as follows: the material entering the reactor in a small-time interval must equal the material leaving the reactor in that interval, plus any accumulation or depletion due to the chemical reaction taking place in the interval. This equation can be expressed as "the elemental time rate of change of the component mass（物料单位时间变化）in the reactor is equal to（等于）the difference between the inlet and outlet mass flows（进出料质量流量差）plus（加）the rate of reaction multiplied by the reactor volume（反应速率乘以反应器体积）."

The energy balance equation can be written similarly，accounting for the rate of heat transfer in the reactor. These equations are typically solved numerically using iterative methods（迭代法）to obtain the reactor design parameters. The design parameters can be optimized for various objectives，such as maximizing yield or minimizing the reactor volume.

技术理论

连续管式反应器内流体的流动处于稳定状态。如图 5-1 所示，没有反应物积累。如图 5-2 所示，沿流体流动方向，物料的浓度、温度和反应速率不断地变化。

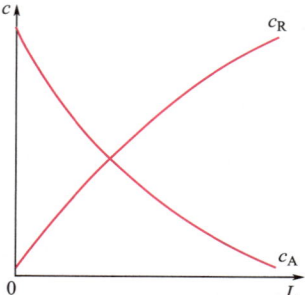

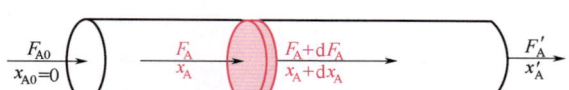

图 5-1　连续管式反应器物料衡算示意

图 5-2　连续管式反应器内物料浓度随管长方向变化关系

因此，以反应物 A 作物料衡算，依据式（2-10）可知

$$dF_A + (-r_A)dV_R = 0 \tag{5-1}$$

简化得

$$V_R = c_{A0}V_0\int_{x_{A0}}^{x_{Af}}\frac{dx_A}{(-r_A)} \tag{5-2}$$

$$\tau = \frac{V_R}{V_0} = c_{A0}\int_{x_{A0}}^{x_{Af}}\frac{dx_A}{(-r_A)} \tag{5-3}$$

式中　τ——物料在连续管式反应器中的空时，h；

V_0——物料进口处体积流量，m^3/h。

关键词详解

空时，airtime，连续管式反应器的有效体积除以物料进口处体积流量，空时是空速（airspeed）的倒数。当反应管内进行的是等容反应时，停留时间与空时相等。

互动练习

5-1 What is the fundamental equation for material balance in a PFR?

A) $H/D=1.2$

B) $V=0.131D^3$

C) Inlet flow rate＝Outlet flow rate＋Reactant consumption rate＋Accumulation rate

D) None of the above

5-2 Which of the following factors is important in designing a PFR?

A) The ratio of height to diameter

B) The volume of the head

C) The shape of the vessel

D) None of the above

5-3 Which of the following methods can be used to determine the number of reactors needed in a series for a given conversion rate?

A) Analytical method

B) Graphical method

C) Numerical method

D) All of the above

5-4 What is the advantage of a PFR over a BR?

A) Higher conversion rates can be achieved

B) The reaction can be run for a longer time

C) The reactor can be shut down easily

D) None of the above

5-5 Which of the following is an important consideration when designing a PFR?

A) The type of reaction being run

B) The size of the reactor

C) The reactor operating temperature

D) All of the above

5.2 Design of Isothermal Constant-Volume PFRs
恒温恒容管式反应器设计

Isothermal constant-volume PFR design is a common method in chemical engineering. The reactor is designed to maintain a constant temperature and volume during the reaction process. This is achieved through a jacketed reactor tube that is surrounded by a temperature-controlled fluid（控温流体）. The design of the reactor requires careful consideration of factors such as the residence time，flow rate，and reaction kinetics. The use of this type of reactor is often preferred for continuous flow reactions as it allows for consistent and precise control of reaction conditions.

5.3 Design of Isothermal Variable-Volume PFRs
恒温变容管式反应器设计

In the design of a constant temperature variable volume（变容）PFR，the volume of the reactor changes during the reaction，which affects the conversion and selectivity of the reactants. The design requires consideration of the thermal expansion coefficient（热膨胀系数）of the reaction mixture and the material of the reactor. The reactor volume can be calculated using the equation of state（状态方程）or the method of initial and final states（初始和终了状态）. The temperature control of the reactor is achieved through heat transfer calculation and proper selection of the cooling or heating medium. The design aims to ensure high reaction efficiency and safety.

5.4 Design of Adiabatic PFRs
绝热连续管式反应器设计

Adiabatic PFRs（绝热连续管式反应器）are designed to maintain a constant temperature without any heat transfer to or from the surroundings. This is achieved by using insulating materials and controlling the reaction temperature through the feed flow rate（进料流量），reactant concentration（反应物浓度），and catalyst loading（催化剂装载）. The design of adiabatic reactors requires careful consideration of factors such as reaction kinetics，heat of reaction，and thermal stability. The reactor size is determined based on the desired conversion rate and the maximum temperature rise during the reaction process. Adiabatic reactors are commonly used in exothermic reactions，and proper safety measures are crucial to prevent thermal runaway and ensure process safety.

技术理论

（1）恒温恒容管式反应器设计

连续管式反应器在恒温恒容过程操作时，可结合恒温恒容条件，计算出达到一定转化率所需要的反应体积或物料在反应器中的停留时间。将动力学方程（如一级不可逆反应或者二级不可逆反应）代入，可求得相应计算反应器体积和转化率的关系。

将物料在间歇釜式反应器的反应时间与在连续管式反应器的停留时间的计算式相比，可以看出在恒温恒容过程时是完全相同的，即在相同的条件下，同一反应达到相同的转化率时，在两种反应器中的时间值相等。这是因为在这两种反应器内，反应物浓度经历了相同的变化过程，只是在间歇釜式反应器内浓度随时间变化，在连续管式反应器内浓度随位置变化而已。也可以说，仅就反应过程而言，两种反应器具有相同的效率，只因间歇釜式反应器存在非生产时间，即辅助时间，故生产能力低于连续管式反应器。

（2）恒温变容管式反应器设计

在反应过程中，因反应温度变化，会发生物料密度的改变，或物料的分子总数改变，导致物料的体积发生变化。通常情况下，液相反应可近似作恒容过程处理，但当反应过程密度变化较大而又要求准确计算时，就要把容积变化考虑进去。对于气相总分子数变化的反应，容积的变化更应考虑。

（3）绝热连续管式反应器设计

在反应进行过程中，系统与外界不发生热量交换的反应器，称为绝热式反应器。

关键词详解

一级不可逆反应，reaction of irreversible one-order，动力学方程式为

$(-r_A)=kc_A$，在恒温条件下 k 为常数，而恒容条件下有 $c_A=c_{A0}(1-x_A)$，并将其代入式（5-3）得：

$$V_R=V_0\tau=c_{A0}V_0\int_{x_{A0}}^{x_{Af}}\frac{\mathrm{d}x_A}{kc_{A0}(1-x_A)}$$

$$=\frac{V_0}{k}\ln\frac{1-x_{A0}}{1-x_{Af}}$$

（5-4）

二级不可逆反应，reaction of irreversible two-order，动力学方程式为 $(-r_A)=kc_A^2$，若 $x_{A0}=0$，同理可得：

$$V_R=V_0\tau=c_{A0}V_0\int_0^{x_{Af}}\frac{\mathrm{d}x_A}{kc_{A0}^2(1-x_A)^2}$$

$$=V_0\frac{x_{Af}}{kc_{A0}(1-x_{Af})}$$

（5-5）

恒温变容管式反应器，isothermal variable-volume PFR，由变容引起的体积、浓度等的变化，可用下述诸式表示：

$$V_t=V_0(1+y_{A0}\delta_A x_A)$$ （5-6）

$$F_t=F_0(1+y_{A0}\delta_A x_A)$$ （5-7）

$$c_A=c_{A0}\frac{1-x_A}{1+y_{A0}\delta_A x_A}$$ （5-8）

$$(-r_A)=-\frac{1}{V}\frac{\mathrm{d}n_A}{\mathrm{d}\tau}=\frac{c_{A0}}{1+y_{A0}\delta_A x_A}\frac{\mathrm{d}x_A}{\mathrm{d}\tau}$$ （5-9）

式中，F_t 为反应系统在操作压力为 p、温度为 T、反应物的转化率为 x_A 时物料的总体积流量，m^3/s。

将以上关系代入反应器基础设计式中，可以求得变容过程反应器有效体积。表 5-1 给出了恒温变容下 $x_{A0}=0$ 时管式反应器设计式。

表 5-1　恒温变容管式反应器计算式

化学反应	速率方程	计算式
A→P（零级）	$(-r_A)=k$	$\dfrac{V_R}{F_{A0}}=\dfrac{x_A}{k_A}$
A→P（一级）	$(-r_A)=kc_A$	$\dfrac{V_R}{F_{A0}}=\dfrac{-(1+\delta_A y_{A0})\ln(1-x_A)-\delta_A y_{A0}x_A}{kc_{A0}}$
2A→P A+B→P （$c_{A0}=c_{B0}$） （二级）	$(-r_A)=kc_A^2$	$\dfrac{V_R}{F_{A0}}=\dfrac{1}{kc_{A0}^2}\left[2\delta_A y_{A0}(1+\delta_A y_{A0})\ln(1-x_A)+\delta_A^2 y_{A0}^2 x_A+(1+\delta_A y_{A0})^2\dfrac{x_A}{1-x_A}\right]$

绝热连续管式反应器，adiabatic plug flow reactors，绝热连续管式反应器的设计计算与前面讨论过的恒温反应器的设计计算方法不同。恒温反应器中反

应速率只是转化率的函数，而绝热反应器的管截面上各点的温度不同，则反应速率不仅是转化率的函数，而且也是温度的函数。所以，须对反应系统列出热量衡算式，然后与物料衡算式、反应动力学方程式联立求解，才能求得为达到一定转化率所需要的反应器有效体积。

由于过程是在绝热条件下进行，所以由系统传递给环境或载热体的热量项为零，又由于是连续操作，系统中热量的积累项也为零，则其热量衡算式为：

$$\begin{bmatrix} \text{微元时间内进入微元体} \\ \text{积的物料所带进的热量} \end{bmatrix} = \begin{bmatrix} \text{微元时间内离开微元体} \\ \text{积的物料带走的热量} \end{bmatrix} - \begin{bmatrix} \text{微元时间微元体积} \\ \text{内由于反应产生的热量} \end{bmatrix}$$

$$F_t'\overline{M}'\overline{c}_p'(T'-T_b)\Delta\tau \qquad F_t\overline{M}\,\overline{c}_p''(T''-T_b)\Delta\tau \qquad (-r_A)(-\Delta H_A)_{T_b}\Delta\tau dV_R$$

即： $F_t'\overline{M}'\overline{c}_p'(T'-T_b)\Delta\tau - F_t\overline{M}\,\overline{c}_p''(T''-T_b)\Delta\tau + (-r_A)(-\Delta H_A)_{T_b}\Delta\tau dV_R = 0$

$$(5\text{-}10)$$

式中　　F_t'——进入微元体积 dV_R 的物料总摩尔流量，kmol/h；

$\qquad\quad F_t$——离开微元体积 dV_R 的物料总摩尔流量，kmol/h；

$\qquad\quad \overline{M}'$——进入微元体积 dV_R 的物料平均相对分子质量，kg/kmol；

$\qquad\quad \overline{M}$——离开微元体积 dV_R 的物料平均相对分子质量，kg/kmol；

$\qquad\quad T'$——进入微元体积 dV_R 的物料温度，K；

$\qquad\quad T''$——离开微元体积 dV_R 的物料温度，K；

$\qquad\quad T_b$——选定的基准温度，K；

$\qquad\quad \overline{c}_p'$——进入微元体积 dV_R 的物料在 $T_b\sim T'$ 温度范围内的平均比热容，
$\qquad\qquad$ kJ/(kg·K)；

$\qquad\quad \overline{c}_p''$——离开微元体积 dV_R 的物料在 $T_b\sim T''$ 温度范围内的平均比热容，
$\qquad\qquad$ kJ/(kg·K)；

$(-\Delta H_A)_{T_b}$——在基准温度下，以反应物 A 计算的化学反应热，kg/kmol。

由于 $F_t'\overline{M}'\overline{c}_p'$ 与 $F_t\overline{M}\,\overline{c}_p$ 在一般情况下差值较小，可以认为它们相等；$(-r_A)dV_R = F_{A0}dx_A$；微元体积 dV_R 内 $T''-T' = dT$，则式（5-10）可简化为：

$$F_t\overline{M}\,\overline{c}_p dT = F_{A0}dx_A(-\Delta H_A)_{T_b} \qquad (5\text{-}11)$$

式（5-11）中，$\overline{M}\,\overline{c}_p$ 是反应混合物组成和温度的函数，$(-\Delta H_A)$ 是温度的函数，变容过程 F_t 又是转化率的函数。故各参数间的函数关系十分复杂，其积分计算也是很麻烦。为了便于计算，可将绝热过程简化为：反应在进口温度 T_0 下恒温进行，使转化率从 x_{A0} 变为 x_{Af}，则化学反应热应取温度在 T_0 时的数值 ΔH_T：

$$\Delta H_T = F_{A0}(x_{Af}-x_{A0})(-\Delta H_A)_{T_0}$$

反应后的混合物由 T_0 恒压升温到温度 T，其升温过程的热量为 ΔH_P，若取 $\overline{c}_p$ 为 $T_0 \sim T$ 范围内的平均值，则

$$\Delta H_P = F_t \overline{M} \overline{c}_p (T_0 - T)$$

对于绝热过程

$$\Delta H = \Delta H_P + \Delta H_T = 0$$

即

$$F_t \overline{M} \overline{c}_p (T - T_0) = F_{A0}(x_{Af} - x_{A0})(-\Delta H_A)_{T_0}$$

故式（5-11）的积分式可写成：

$$T - T_0 = \frac{F_{A0}(-\Delta H_A)_{T_0}}{F_t \overline{M} \overline{c}_p}(x_{Af} - x_{A0}) \tag{5-12}$$

式（5-12）为绝热过程连续管式反应器内温度与转化率之间的函数关系式，$(T - T_0)$ 为达到出口转化率 x_{Af} 时反应器的最大温差。将式（5-2）代入式（5-2），可用以计算绝热式连续管式反应器为达到一定转化率所需要的有效体积，或物料在反应器中的停留时间。

如果反应过程无摩尔数的变化，即 $F_t = F_0$，又 $x_{A0} = 0$，其他各参数取值基准仍与以上简化方案相同，则式（5-12）可写为：

$$T - T_0 = \frac{F_{A0}(-\Delta H_A)_{T_0}}{F_0 \overline{M} \overline{c}_p} x_{Af}$$

或

$$x_{Af} = \frac{F_0}{F_{A0}} \cdot \frac{\overline{M} \overline{c}_p (T - T_0)}{(-\Delta H_A)_{T_0}} = \frac{\overline{M} \overline{c}_p (T - T_0)}{y_{A0}(-\Delta H_A)_{T_0}} \tag{5-13}$$

互动练习

5-6　What is the purpose of using a jacketed reactor tube in a constant temperature and volume PFR design?

A）To maintain a constant flow rate

B）To control reaction kinetics

C）To maintain a constant temperature and volume during the reaction process

D）To prevent thermal runaway

5-7　What is the main difference between a constant temperature variable volume PFR and a constant temperature and volume PFR?

A）The former allows for a change in volume during the reaction process

B）The former maintains a constant temperature and volume during the reaction process

C）The latter is used for exothermic reactions only

D）The latter does not require consideration of the thermal expansion coefficient of the reaction mixture

5-8　What is the purpose of using insulating materials in the design of an adiabatic PFR?

A）To maintain a constant temperature

B）To control reaction kinetics

C）To prevent thermal runaway

D）To reduce the volume of the reactor

5-9　What factors need to be considered in the design of adiabatic continuous PFRs?

A）Reactor size and catalyst loading

B）Thermal stability and heat of reaction

C）Residence time and flow rate

D）Reactor volume and conversion rate

5-10　In which type of reaction are adiabatic PFRs commonly used?

A）Endothermic reactions

B）Exothermic reactions

C）Hydrolysis reactions

D）Reduction reactions

任务 6

Design and Operation Optimization of Homogeneous Reactors
均相反应器设计与操作优化

任务要点

本任务讲解均相反应器的设计和操作优化方法，读者将学习如何通过调整温度、压力、流速等参数优化反应器性能，并掌握降低能耗、提高生产效率的具体措施，为清洁化工生产提供技术支持。

学习目标

知识目标

（1）了解均相反应器的设计原则及优化方法。

（2）掌握操作条件对均相反应器性能的影响。

技能目标

（1）能根据实际需求优化均相反应器的设计与操作参数。

（2）能利用实验数据改进反应器的性能。

价值目标

（1）强调高效生产与节能的结合。

（2）培养问题分析和解决的能力。

6.1 Comparison of Production Capacities of Reactors for Simple Chemical Reactions
简单反应的反应器生产能力比较

When it comes to the production of simple chemical reactions, the choice of reactor type can have a significant impact on the productivity and efficiency of the process. The most used reactors for simple reactions are BRs and CSTRs, as well as plug flow reactors (PFRs). The choice of reactor depends on factors such as the reaction kinetics，the desired conversion rate，and the reaction heat transfer requirements.

In general，PFRs are preferred over CSTRs for reactions with a high conversion rate（高转化率），as they offer better plug flow characteristics and reduced mixing effects（混合作用）. However，for reactions with low conversion rates（低转化率），CSTRs may be more efficient due to their better temperature control and mixing capabilities.

In terms of continuous reactor types，CSTRs and PFRs are commonly used for simple reactions. PFRs offer a more efficient use of space and better heat transfer，but they may suffer from mass transfer limitations（传质限制）and pressure drop issues. CSTRs，on the other hand，provide better mixing and easier control of reaction conditions.

For reactions that require multiple reactors in series，multiple sequential CSTRs are often used due to their ease of control and ability to handle large volumes. However，Multiple parallel PFRs in series can also be used to achieve better conversion rates.

In some cases，combination reactors（组合反应器）can be used to optimize reaction efficiency. For example，a self-catalyzed reaction（自催化反应）can be optimized using a combination of CSTRs and PFRs，where the CSTR provides the initial reaction and the PFR provides further reaction optimization.

Overall，the choice of reactor type and combination depends on the specific requirements of the reaction and process，and careful consideration must be taken to ensure optimal performance and efficiency.

技术理论

简单反应的反应器组合与优化包括单个反应器间的比较、多釜串联连续釜式反应器的比较以及组合反应器的优化。它涉及反应器设计、操作参数的调整和流程优化，旨在提高反应效率、减少能源消耗、优化产物纯度等方面取得最佳结果。

主要步骤和考虑因素包括：

• 反应器选择：根据反应类型、动力学特性、热效应等，选择适当类型的反应器，如批式反应器、连续流动反应器、循环流化床反应器等。

- 反应组合：将不同类型的反应器进行组合，例如串联多个反应器以在不同阶段进行不同的反应步骤，或并联多个反应器以平行处理多个反应。

- 热管理：设计和优化反应器的冷却系统，以控制温度，避免温度失控和副反应的发生。

- 催化剂与催化策略：优化催化剂的选择和使用，考虑催化剂的寿命、再生等因素，以提高催化效率。

- 流程控制：确定最佳的操作参数，如温度、压力、流量等，以实现最佳的反应条件。

- 分离与纯化：考虑反应产物的分离和纯化，设计合适的分离单元，以提高产物纯度和收率。

- 能源效率：优化能源利用，考虑热集成和废热回收等方式，减少能源消耗。

- 安全性与环保：确保化学过程的安全性，预防事故发生，同时考虑环保因素，减少废物排放和环境影响。

（1）单个反应器间的比较

1）间歇釜式反应器和连续管式反应器的比较

对间歇操作反应器，其反应时间为：

$$\tau_m = c_{A0} \int_0^{x_{Af}} \frac{dx_A}{(-r_A)} \tag{6-1}$$

式中，τ_m 为间歇釜式反应器的反应时间，h。

对连续管式反应器

$$\tau_p = \frac{V_{Rp}}{V_0} = c_{A0} \int_0^{x_{Af}} \frac{dx_A}{(-r_A)} \tag{6-2}$$

式中，τ_p 为连续管式反应器的反应时间，h；V_{Rp} 为连续管式反应器的有效体积，m^3。

由式(6-1) 和式(6-2) 可知，$\tau_m = \tau_p$。仅从反应时间而言，在间歇釜式反应器和连续管式反应器中进行时，所需反应时间是相同的。但由于间歇操作需要辅助时间，所以实际计算时不能以反应时间为准，而以操作周期 $\tau_m + \tau_辅$ 为准，需要的反应器体积比连续管式反应器的体积要大。连续管式反应器不存在辅助时间，也没有装料系数问题。

2）连续釜式反应器和连续管式反应器的比较

对连续釜式反应器：

$$V_{Rc} = \frac{V_0 c_{A0} x_{Af}}{(-r_A)} = \frac{F_{A0} x_{Af}}{(-r_A)}$$

或
$$\tau_c = \frac{V_{Rc}}{V_0} = \frac{V_{Rc}c_{A0}}{F_{A0}} = \frac{c_{A0}x_{Af}}{(-r_A)} \tag{6-3}$$

则
$$\frac{\tau_c}{\tau_p} = \frac{V_{Rc}}{V_{Rp}} = \frac{\dfrac{x_{Af}}{(-r_A)}}{\displaystyle\int_0^{x_{Af}} \frac{\mathrm{d}x_A}{(-r_A)}} \tag{6-4}$$

式中，V_{Rc} 为连续釜式反应器的有效体积，m^3；τ_c 为连续釜式反应器的反应时间，h。

将反应速率和具体操作条件代入式（6-4）便可计算使用两种型式反应器的有效体积大小比较关系。如恒容恒温过程的幂指数型动力学方程式为 $(-r_A) = kc_A^n$，有

$$\frac{\tau_c}{\tau_p} = \frac{V_{Rc}}{V_{Rp}} = \frac{(n-1)x_{Af}}{(1-x_{Af}) - (1-x_{Af})^n}, \quad n \neq 1 \tag{6-5}$$

或
$$\frac{\tau_c}{\tau_p} = \frac{V_{Rc}}{V_{Rp}} = \frac{\dfrac{x_{Af}}{(1-x_{Af})}}{-\ln(1-x_{Af})} = \frac{x_{Af}}{(x_{Af}-1)\ln(1-x_{Af})}, \quad n = 1 \tag{6-6}$$

以式（6-5）和式（6-6）用对比时间和对比体积对 n、x_{Af} 作图，即可看到有效体积比随着不同反应达到不同转化率时的变化关系，如图 6-1 所示。

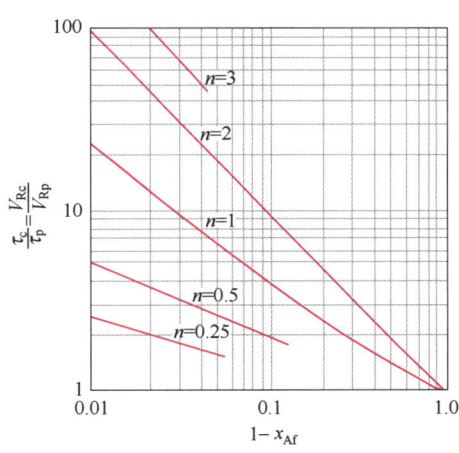

图 6-1 n 级反应在恒温恒容单个反应器中的性能比较

由图 6-1 可以看出，当转化率很小时，反应器的性能受流动状态的影响较小，当转化率趋于 0 时，连续釜式反应器与连续管式反应器体积比等于 1，即 $V_{Rc} = V_{Rp}$，$\tau_c = \tau_p$。而随着转化率的增加，两者体积比相差愈来愈显著。由此得出这样的结论：过程要求进行的程度（转化率）越高，返混影响就越大，因此对高转化率的反应，宜采用连续管式反应器。

（2）多釜串联连续釜式反应器的比较

从连续釜式反应器和连续管式反应器的计算公式

$$\tau_p = \frac{V_{Rp}}{V_0} = c_{A0} \int_0^{x_{Af}} \frac{dx_A}{(-r_A)} \text{ 和 } \tau_{ci} = \frac{V_{Rci}}{V_0} = \frac{c_{A0}(x_{Ai} - x_{Ai-1})}{(-r_A)_i}$$

出发，对同一反应达到同样的转化率，可以以图 6-2 的形式表明两种反应器的体积比。

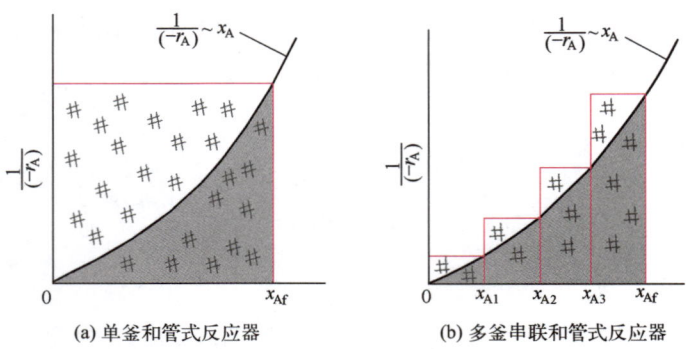

(a) 单釜和管式反应器   (b) 多釜串联和管式反应器

图 6-2 理想混合反应器和理想置换反应器体积比较

图 6-2 中（a）为单台连续釜式反应器和连续管式反应器体积之比的关系。图中矩形面积为 τ_c/c_{A0}，曲线下面的积分面积为 τ_p/c_{A0}，很显然，$\tau_c > \tau_p$，即 $V_{Rc} > V_{Rp}$，即单只连续釜式反应器的体积大于连续管式反应器的有效体积。

图 6-2 中（b）为同一反应达到同样的转化率使用多台串联连续釜式反应器和连续管式反应器的比较。按下式

$$\tau_{ci} = \frac{V_{ci}}{V_0} = \frac{c_{A0}(x_{Ai} - x_{Ai-1})}{(-r_A)_i}$$

可得各个小矩形面积为 $\tau_{ci}/c_{A0} = \Delta x_{Ai}/(-r_A)_i$，其总面积之和要比单釜时的大矩形面积要小得多，且串联釜数越多，需总反应器的体积越小。当串联釜数无限多时，则和连续管式反应器体积相同。因为每釜之间没有返混，从最前面一釜开始，各釜中的反应物浓度和反应速率由高到低，最后达到要求的转化率，这就是生产中为何采用多釜串联反应器的主要原因之一。

（3）组合反应器的优化

前面介绍了在多台体积相同的连续釜式反应器串联时，完成同一个反应 τ_c/τ_p 值随着釜数的增加而减少，即总有效体积 V_{Rc} 变小。如果使用同样的釜数串联，达到相同的最终转化率，在各釜大小不同时，则其总需有效体积是不同的，因此有必要讨论有关多釜串联连续釜式反应器组合的优化问题。

1）多釜串联连续釜式反应器的优化

不同大小的多只连续釜式反应器串联操作时，若最终转化率已经给定，如何确定其最优组合？先来介绍只有两只反应釜串联的情况。

图 6-3 表示的关系是两个反应器的交替排列，两者都达到相同的最终转化率，设法使体积最小，应选最优的 x_{A1}，也就是确定图上 B 点的位置，使矩形 $ABCD$ 的面积最大。只有当 B 点正好处于曲线上斜率等于矩形对角线 AC 的斜率时矩形面积为最大。一般来说，对于 $n>0$ 的幂指数函数的动力学时，总是正好有一个"最优点"，如图 6-4 所示。

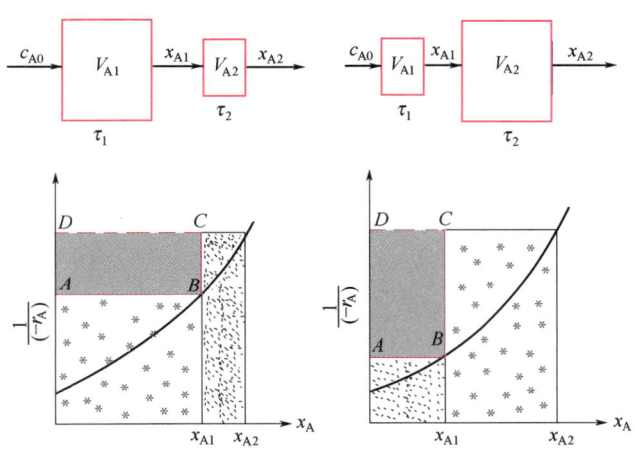

图 6-3　不同大小双釜串联比较

对于"最优点" x_{A1}，也可用计算法直接求取。按多只串联连续釜式反应器计算公式得

$$\tau_1 = \frac{c_{A0} x_{A1}}{(-r_A)_1}$$

$$\tau_2 = \frac{c_{A0}(x_{A2} - x_{A1})}{(-r_A)_2}$$

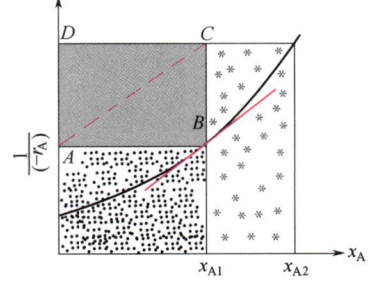

图 6-4　矩形面积法求最优
中间转化率

当两釜串联时，两釜中的总停留时间等于两釜各自停留时间之和，即

$$\tau = \tau_1 + \tau_2 = \frac{c_{A0} x_{A1}}{(-r_{A1})} + \frac{c_{A0}(x_{A2} - x_{A1})}{(-r_{A2})} = \frac{c_{A0} x_{A1}}{k_1 f(x_{A1})} + \frac{c_{A0}(x_{A2} - x_{A1})}{k_2 f(x_{A2})}$$

对于两釜串联中进行一级不可逆反应，且两釜反应温度相同时，令 $\dfrac{\mathrm{d}\tau}{\mathrm{d}x_{A1}} = 0$，得

$$x_{A1} = 1 - (1 - x_{A2})^{1/2}$$

可见，对于一级反应，各釜大小相同时是最优的。对于反应级数 $n \neq 1$，$n > 0$，较小的反应器在前面，而对于 $n < 0$ 应先用较大的反应器。不同的情况应具体分析计算。

2）自催化反应过程的优化

自催化反应表示为 $A + P \rightarrow P + P$，其反应速率方程为

$$(-r_A) = kc_A c_p \tag{6-7}$$

反应的产物本身具有催化作用，加速了反应的继续进行，形成一个正反馈循环。自催化反应通常表现为初始阶段反应速率较慢，但随着产物的生成，反应速率逐渐加快。这种反应机制可以导致反应过程迅速增加，直至达到一定平衡或最终产物。自催化反应在如生化反应的发酵、废水生化处理等方面都有广泛的应用。

严格地讲，对于自催化反应，如果原料中不存在产物时，反应速率应为零，反应不能进行，通常情况下则将少量反应产物加入原料中。

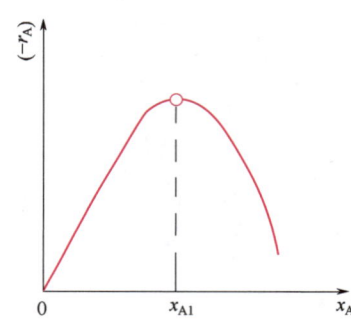

图 6-5　自催化反应速率规律示意图

在反应初期，虽然反应物 A 的浓度高，但此时作为催化剂的反应产物 P 的浓度很低，所以反应速率较低。随着反应的进行，反应产物 P 的浓度逐渐增加，反应速率加快。在反应后期，虽然产物 P 的浓度很高，但因反应物 A 的消耗，其浓度大大降低，此时反应速率又下降。由此可见，自催化反应过程的基本特征是存在一个最大反应速率点，如图 6-5 所示。自催化反应虽然有其独特的反应速率特征，但它在反应器中反应结果仍然可以用简单反应的处理方法进行计算。

根据自催化反应存在最大反应速率点的特征，在反应器选型时，根据不同转化率的要求，选用不同的反应器及其组合型式，以减小反应器体积。下面以图解法进行讨论。如图 6-6 所示，以 $x_A \sim 1/(-r_A)$ 作图。如果自催化反应所要求转化率小于或等于 x_{A1}，如图 6-6(c) 所示，为达到相同转化率，连续釜式反应器显然比连续管式反应器体积要小，表明返混是有利因素。因为返混导致反应器内产物和原料相混合，使低转化率时反应器内也有较高的产物浓度，得到较高的反应速率。相反，当要求最终转化率较高时，如图 6-6(a) 所示，返混则导致整个反应器处于低的原料浓度，反应速率很低，所以为达到相同转化率，连续釜式反应器所需体积将大于连续管式反应器。当反应处于中等转化率时。如图 6-6(b) 所示，则两类反应器无多大差别。

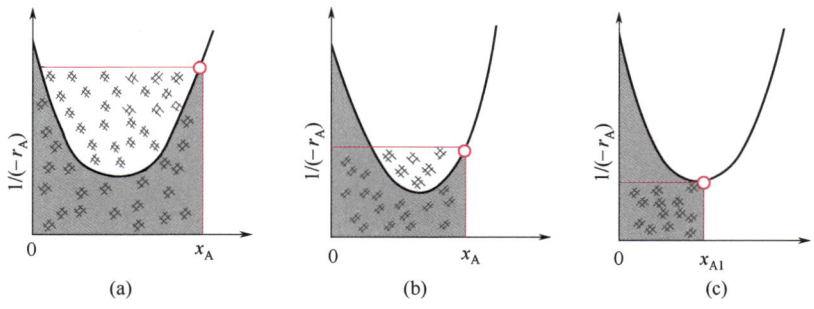

图 6-6　连续釜式反应器和连续管式反应器用于自催化反应性能比较

为了使反应器总体积最小，可选用一个连续釜式反应器，使反应器保持在最高速率点处进行反应。为了使反应原料得到充分利用，达到较高的转化率，可以在连续釜式反应器后串联一个连续管式反应器来达到高转化率要求。这里的最优反应器组合是先用一个连续釜式反应器，控制在最大速率点处操作，然后接一个连续管式反应器，达到高转化率以充分利用原料，其组合如图 6-7（a）所示。也可以在连续釜式反应器出口接一个分离装置，将反应器出口物料分离产物后原料返回反应器。其最优组合为一个连续釜式反应器后接一个分离装置，连续釜式反应器控制在最大速率点处操作，如图 6-7（b）所示。

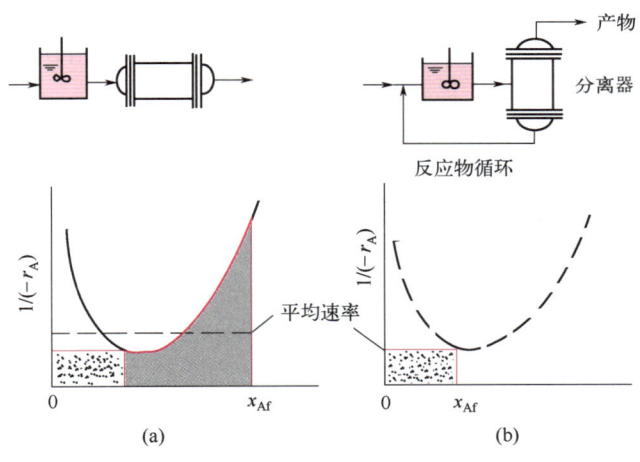

图 6-7　反应器组合的最优化

反应器的组合与优化，combination and optimization of chemical reactors，即将不同类型的反应器以及操作策略组合在一起，以实现最佳的反应条件和产品产量。

自催化反应，autocatalytic reaction，指在化学反应中，反应生成物作为反应的催化剂，从而促进同一反应的进行。

互动练习

6-1　For a reaction with a high conversion rate，which type of reactor is preferred?

A）BR

B）CSTR

C）PFR

D）Both B and C

6-2　Which type of reactor is more efficient for reactions with low conversion rates?

A）BR

B）CSTR

C）PFR

D）Both B and C

6-3　What are the advantages of using a PFR over a CSTR?

A）Better mixing and easier control of reaction conditions

B）More efficient use of space and better heat transfer

C）Reduced mixing effects and better plug flow characteristics

D）BothA and B

6-4　Which type of reactor is often used for reactions that require multiple reactors in series?

A）BR

B）CSTR

C）PFR

D）Both B and C

6-5　How can a self-catalyzed reaction be optimized using a combination of reactors?

A）Using a BR and a PFR

B）Using multiple parallel PFRs

C）Using a CSTR and a PFR

D）Using a PFR and another PFR

6.2 Selectivity Comparison of Complex Chemical Reactions
复杂反应的选择性比较

Selectivity comparison of complex reactions is an essential aspect of chemical process engineering. In many cases，complex reactions involve parallel reactions，where two or more reactions occur simultaneously，leading to the formation of different products. The selectivity of a reaction is the ability of the reaction to produce a specific product with high yield and purity.

Parallel reactions can be classified into two types：reactions involving one reactant that forms a primary product（主产物）and a secondary product（副产物）and reactions involving two reactants that form a primary product and a secondary product. In the former，the selectivity of the reaction is determined by the reaction rate and the relative stabilities（相对稳定度）of the primary and secondary products. In the latter，the selectivity of the reaction is influenced by factors such as the reaction rate，the relative stabilities of the products，and the ratio of the two reactants（两个反应物的比率）.

Composite complex reactions involve a series of elementary reactions（一系列的基元反应），where each step is reversible（可逆的）and can occur in multiple ways，leading to different products. The selectivity of composite reactions depends on factors such as the reaction rate，the relative stabilities of the intermediates（中间产物），and the activation energies of the individual steps（单个步骤）.

Selective comparison of complex reactions requires a thorough understanding of the reaction mechanism（对反应机理的充分理解），the kinetics，and the thermodynamics（热力学）of the reaction. Experimental techniques such as spectroscopy（光谱学），chromatography（色谱学），and mass spectrometry（质谱学）can be used to identify the products and intermediates of the reaction and determine their concentrations. Mathematical models can be developed to simulate the reaction and predict the selectivity of the reaction under different conditions.

In summary，the selective comparison of complex reactions is a critical aspect of chemical process engineering，involving the understanding of parallel reactions，composite complex reactions，and the experimental and mathematical techniques to identify and predict the selectivity of the reaction.

技术理论

复杂反应的种类很多，其基本反应是平行反应和连串反应，由平行反应和连串反应形成复合复杂反应。在选择反应器型式和操作方法时，对复杂反应过程必须考虑反应的选择性问题。

关键词详解

平行反应，parallel reaction，平行反应一般包括以下情况。

$$
A \Big\langle\begin{array}{l} \xrightarrow{k_1} \text{R 主产物} \\ \xrightarrow{k_2} \text{S 副产物} \end{array}
$$

（1）反应为一种反应物生成一种主产物和一种副产物

此类平行反应得到较多目的产物 R 所应采用的反应器类型和操作方式，可通过动力学分析。它们的反应动力学方程式为：

$$r_R = \frac{dc_R}{d\tau} = k_1 c_A^{\alpha_1}$$

$$r_S = \frac{dc_S}{d\tau} = k_2 c_A^{\alpha_2}$$

定义选择性
$$S_P = \frac{r_R}{r_S} = \frac{k_1}{k_2} c_A^{\alpha_1 - \alpha_2} \tag{6-8}$$

可见，增大 r_R/r_S 可以增大反应的选择性，亦即得到较多的 R。因为在一定反应系统和温度时，k_1、k_2、α_1、α_2 均为常数。故只要调节反应物浓度 c_A，就可得到较大的 r_R/r_S 值。由式(6-8) 可得以下结论。

① 当 $\alpha_1 > \alpha_2$ 时，提高反应物浓度 c_A 则可使 r_R/r_S 增大。因为连续管式反应器内反应物的浓度较连续釜式反应器为高，故适宜于采用连续管式反应器，次则采用间歇釜式反应器或连续多釜串联反应器。

② 当 $\alpha_1 < \alpha_2$ 时，降低反应物浓度 c_A 则可使 r_R/r_S 增大。为此，适宜于采用连续釜式反应器。但在完成相同生产任务时，所需釜式反应器体积较大。故需全面分析，再作选择。

③ $\alpha_1 = \alpha_2$ 时，$S_P = \dfrac{r_R}{r_S} = \dfrac{k_1}{k_2} = $ 常数，则反应物浓度的改变对选择性无影响。

（2）反应为两种反应物生成一种主产物和一种副产物

$$A + B \xrightarrow{k_1} R \text{ 主产物}, \qquad A + B \xrightarrow{k_2} S \text{ 副产物}$$

它们的动力学方程式分别为：

$$r_R = k_1 c_A^{\alpha_1} c_B^{\beta_1}$$

$$r_S = k_2 c_A^{\alpha_2} c_B^{\beta_2}$$

则反应的选择性 S_P 为：

$$S_P = \frac{r_R}{r_S} = \frac{k_1}{k_2} c_A^{\alpha_1 - \alpha_2} c_B^{\beta_1 - \beta_2} \tag{6-9}$$

为了使选择性亦即 r_R/r_S 比值为最大，对各种所希望的反应物浓度的高、低或高-低结合，完全取决于竞争反应的动力学。这些浓度的控制，可以按进料方式和反应器类型而调整。表 6-1 和表 6-2 表示了存在两个反应物的平行反应在间歇和连续操作时保持竞争浓度使之适应竞争反应动力学要求的情况。

表 6-1　间歇操作时不同竞争反应动力学下的操作方式

动力学特点	$\alpha_1 > \alpha_2$，$\beta_1 > \beta_2$	$\alpha_1 < \alpha_2$，$\beta_1 < \beta_2$	$\alpha_1 > \alpha_2$，$\beta_1 < \beta_2$
控制浓度要求	应使 c_A、c_B 都高	应使 c_A、c_B 都低	应使 c_A 高、c_B 低
操作示意图			
加料方法	瞬间加入所有的 A 和 B	缓缓加入 A 和 B	先把全部 A 加入，然后缓缓加 B

表 6-2　连续操作时不同竞争反应动力学下的操作方式其浓度分布

动力学特点	$\alpha_1 > \alpha_2$，$\beta_1 > \beta_2$	$\alpha_1 < \alpha_2$，$\beta_1 < \beta_2$	$\alpha_1 > \alpha_2$，$\beta_1 < \beta_2$
控制浓度要求	应使 c_A、c_B 都高	应使 c_A、c_B 都低	应使 c_A 高、c_B 低
操作示意图			
浓度分布图			

连串反应，sequential reaction，相比平行反应其情况更为复杂，在此只讨论一级反应。对于连串反应：

$$A \xrightarrow{k_1} R \xrightarrow{k_2} S$$

它们的动力学方程为：

$$r_R = \frac{dc_R}{d\tau} = k_1 c_A - k_2 c_R$$

$$r_S = \frac{dc_S}{d\tau} = k_2 c_R$$

则反应的选择性 S_P 为：

$$S_P = \frac{r_R}{r_S} = \frac{k_1 c_A - k_2 c_R}{k_2 c_R} \tag{6-10}$$

由式(6-10) 可知：如 R 为目的产物，当 k_1、k_2 一定时，为使选择性 S_P 提高，即为使 r_R/r_S 比值增大，应使 c_A 高 c_R 低，适宜于采用连续管式反应器、间歇釜式反应器和连续多釜串联反应器。反之，若 S 为目的产物，则应 c_A 低 c_R 高，适宜于采用连续釜式反应器。但应注意，连串反应 R 生成的增加，有利于 S 的生成（特别是 $k_1 \ll k_2$ 时）的特点，故以 R 为目的产物时，应保持较低的单程转化率。当 $k_1 \gg k_2$ 时，可保持较高的反应转化率，这样可使选择性降低较少，但反应后的分离负荷却可以大为减轻，如图 6-8 所示。

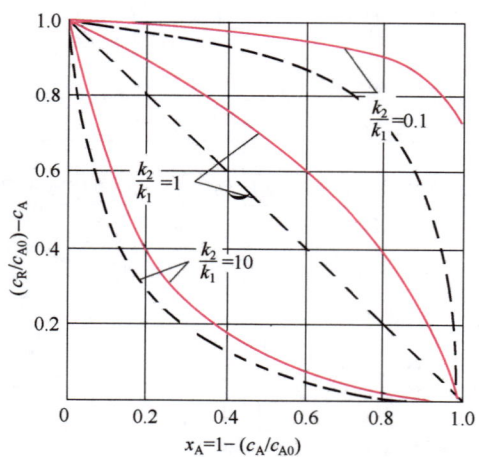

图 6-8　连续管式和釜式反应器选择性比较

图中——连续管式反应器；－－连续釜式反应器

由图 6-8 可以看到：

① 连续操作管式反应器的选择性高于连续釜式反应器；

② 连串反应的选择性随反应转化率的增大而下降；

③ 选择性与速率常数比值 k_2/k_1 密切相关，比值 k_2/k_1 越大，其选择性随转化率的增加而下降的趋势越严重。

根据以上分析可以知道，连串反应转化率的控制十分重要，不能盲目追求反应的高转化率。在工业生产上经常使反应在低转化率下操作，以获得较高的选择性。而把未反应的原料经分离后返回反应器循环使用，此时应以反应-分离系统的优化经济目标来确定最适宜的反应转化率。

复合复杂反应，complex compound reaction，如下所示的反应

$$A+B \xrightarrow{k_1} R$$

$$R+B \xrightarrow{k_2} S$$

$$A \xrightarrow{k_3} R \xrightarrow{k_4} S$$

即为典型的复合复杂反应。此反应中，对 B 而言是平行反应，对 A、R、S 而言则为连串反应。在处理复合复杂反应时，应根据具体情况分别处理。如解决 B 的转化率为主时，把复合复杂反应以平行反应处理；如果以 A 的转化率为主时，以连串反应处理。

互动练习

6-6　What is the selectivity of a reaction?

A）The ability of the reaction to produce a specific product with high yield and purity

B）The number of reactants involved in the reaction

C）The time it takes for the reaction to occur

D）The concentration of the reactants and products

6-7　What are the two types of parallel reactions?

A）Reactions involving one reactant and reactions involving two reactants

B）Reactions involving high-energy intermediates and reactions involving low-energy intermediates

C）Reactions involving acid and base catalysts and reactions involving no catalyst

D）Reactions involving light and reactions not involving light

6-8　What are composite complex reactions?

A）Reactions involving one reactant and multiple products

B）Reactions involving two reactants and multiple products

C）Reactions involving a series of elementary reactions

D）Reactions involving a single elementary step

6-9　Which of the following experimental techniques is NOT used to identify the products and intermediates of a complex reaction?

A）Spectroscopy

B）Chromatography

C）Mass spectrometry

D）Electrolysis

6-10　What is the role of mathematical models in the selective comparison of complex reactions?

A）To simulate the reaction and predict the selectivity of the reaction under different conditions

B）To identify the products and intermediates of the reaction

C）To measure the activation energy of the reaction

D）To determine the concentration of the reactants and products

任务7

Operation and Control of Atmospheric BRs
常压间歇釜式反应器操作与控制

任务要点

本任务介绍常压间歇釜式反应器的基本操作与控制方法，包括反应器运行参数的调控及常见问题的解决方案。读者将学会如何制定科学的操作流程，提高生产效率和产品质量，确保生产过程安全。

学习目标

知识目标

（1）掌握常压间歇釜式反应器的基本操作方法。

（2）了解常见操作问题及其解决方案。

技能目标

（1）能制定并执行常压反应器的操作计划。

（2）能监控操作参数以提高生产效率。

价值目标

（1）注重操作过程中的安全管理。

（2）提高操作人员的团队协作能力。

7.1 Process Overview of 2-Mercaptobenzothiazole
2-巯基苯并噻唑工艺流程简述

2-Mercaptobenzothiazole（MBT）（2-巯基苯并噻唑）is an important organic intermediate used in the production of rubber chemicals and other organic compounds. The production process of MBT involves the reaction of 2-mercaptobenzothiazole with aniline in the presence of a catalyst. The reaction principle is based on the formation of a complex intermediate that undergoes cyclization

（成环）and dehydrogenation（脱氢）to form MBT.

The process flow of MBT production involves several steps，including the preparation of the reactants，the reaction step，and the separation and purification of the product. The reactants，2-mercaptobenzothiazole and aniline，are first mixed and heated in the presence of a catalyst to initiate the reaction（开始反应）. The reaction mixture is then subjected to a series of separation and purification steps（分离和纯化步骤），such as distillation（精馏），extraction（萃取），and crystallization（结晶），to obtain the final product of MBT.

7.2 Operating and Control of Atmospheric BRs for MBT 间歇操作釜式 MBT 反应器操作与控制

The production of MBT in a atmospheric BR involves several steps，including start-up（开车），normal operation（正常操作），and shutting down（停车）.

Before starting the reactor，the raw materials should be properly weighed and prepared to ensure the accuracy of the reaction.

During the reaction，it is important to control the process parameters（工艺参数），such as temperature，pressure，and stirring speed，to achieve optimal product yield and quality. The operator should also monitor the reaction progress and adjust the production parameters accordingly.

After the reaction is complete，the product should be discharged，and the reactor should be cleaned thoroughly before the next batch. In case of emergencies（紧急情况）or abnormal situations（不正常情况），appropriate measures should be taken to ensure the safety of the personnel（全体员工）and the equipment. Proper operation and control of the BRs is essential for the efficient and safe production of MBT.

7.3 Common Abnormal Phenomena and Solutions in MBT Production Reactors MBT 生产反应器中的不正常现象与解决方案

During the production of MBT in a reactor，several abnormal phenomena

（不正常现象）may occur，such as excessive foaming（过度起泡），overheating（过热），and agglomeration of solid particles（固体颗粒结块）.

These anomalies may affect the reaction efficiency and the quality of the product，and therefore require appropriate measures to be taken. Excessive foaming can be controlled by adjusting the stirring speed or adding defoamers（消泡剂）. Overheating can be prevented by adjusting the cooling system or reducing the feed rate. Agglomeration of solid particles can be avoided by optimizing the mixing conditions or adding dispersants（分散剂）. In case of emergencies or unexpected situations，appropriate emergency measures should be taken to ensure the safety of the personnel and the equipment. Proper identification and timely handling of abnormal phenomena is crucial for the efficient and safe production of MBT.

技术理论

2-巯基苯并噻唑生产工艺流程如图 7-1 所示，来自备料工序的 CS_2、$C_6H_4ClNO_2$、Na_2S_n 分别注入计量罐及沉淀罐中，经计量沉淀后利用位差及离心泵压入反应釜中，釜温由夹套中的蒸汽、冷却水及蛇管中的冷却水控制，通过控制反应釜温度来控制反应速率及副反应速率，以获得较高的收率及确保反应过程安全。

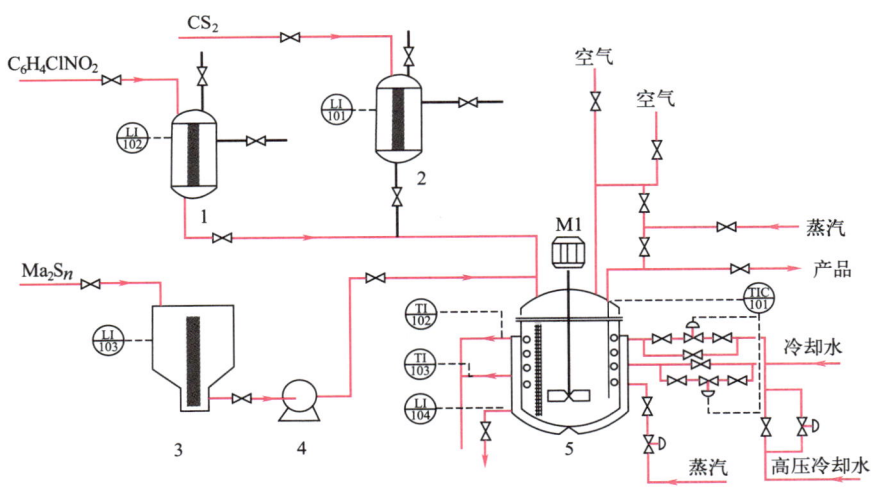

图 7-1　2-巯基苯并噻唑工艺流程图

1—邻硝基氯苯计量罐；2—二硫化碳计量罐；3—多硫化钠沉淀罐；

4—离心泵；5—间歇釜式反应器

关键词详解

开车，start-up，反应设备的开车是指启动、启用或恢复运行反应设备，以便进行特定的化学、物理或工艺反应。这通常涉及将设备的各个组件和系统逐步启动，确保其达到所需的操作参数和条件，以便开始所需的反应过程。在开车过程中，操作人员可能需要监测设备的各种指标、参数和信号，确保设备安全稳定地运行，并在需要时进行调整和干预。开车过程通常需要遵循预定的操作程序和安全规范，以确保反应设备在操作中达到预期的效果，并最大程度地减少潜在的风险或意外事件。

本工艺开车之前还要进行备料和进料，本工艺的开车过程即指开启反应釜搅拌电机、适当打开夹套蒸汽加热阀，观察反应釜内温度和压力上升情况，控制适当的升温速度，逐渐使反应温度、压力等工艺指标达到正常值并进行稳定运行。

正常运行，normal operation，反应器的正常运行是指使反应器按照设定的工艺参数要求进行运行。本工艺的主要工艺生产指标的调整方法包括：

• 温度调节：操作过程中以温度为主要调节对象，以压力为辅助调节对象。升温慢会引起副反应速率大于主反应速率的时间段过长，因而反应的产率低；升温快则容易反应失控。

• 压力调节：压力调节主要是通过调节温度实现的，但在超温时可以微开放空阀，使压力降低，以达到安全生产的目的。

• 收率调节：由于在90℃以下时副反应速率大于正反应速率，因此在安全的前提下快速升温是高收率的保证。

停车，shutting down，反应设备的停车是指将正在进行的化学、物理或工艺反应的设备暂时停止运行或关闭，以便进行维护、维修、更换部件、调整操作条件、处理故障或进行其他必要的操作。这种停车可以是计划性的，也可以是紧急性的，目的是确保设备的正常运行、安全性和效率，以及防止潜在的操作风险或事故。在停车过程中，操作人员可能需要遵循特定的程序和安全准则，以确保设备在停车期间得到适当的处理，并在停车后能够重新启动并继续进行反应。停车操作通常需要经过精心的计划和协调，以确保对生产过程的最小干扰。本工艺在冷却水量很小的情况下，反应釜的温度下降仍较快，则说明反应接近尾声，可以进行停车出料操作。其过程如下：

① 打开反应釜放空阀，放掉釜内残存的可燃气体，然后关闭放空阀。

② 打开蒸汽总阀，打开蒸汽加压阀给釜内升压，使釜内气压高于 4atm。

③ 打开蒸汽预热阀片刻。

④ 打开反应釜出料阀门出料，出料完毕后进行吹扫，然后关闭出料阀，关闭蒸汽阀。

互动练习

7-1　What is 2-mercaptobenzothiazole（MBT）used for?

A）Production of agricultural chemicals

B）Production of rubber chemicals

C）Production of pharmaceuticals

D）Production of food additives

7-2　What is the reaction principle for the production of MBT?

A）Hydrolysis of aniline

B）Oxidation of 2-mercaptobenzothiazole

C）Cyclization and dehydrogenation of a complex intermediate

D）Reduction of a catalyst

7-3　What are the important process parameters for MBT production?

A）pH and salinity

B）Temperature and pressure

C）Particle size and distribution

D）Flow rate and residence time

7-4　What are the steps involved in the production of MBT?

A）Charging，reaction，and discharging

B）Filtration，drying，and packaging

C）Extraction，distillation，and crystallization

D）Hydrolysis，reduction，and oxidation

7-5　What are the solutions for common abnormal phenomena in the reactor during MBT production?

A）Adjusting the stirring speed or adding defoamers for excessive foaming

B）Adjusting the cooling system or reducing the feed rate for overheating

C）Optimizing the mixing conditions or adding dispersants for agglomeration of solid particles

D）All of the above

Operation and Control of High-Pressure BRs
高压间歇釜式反应器操作与控制

任务要点

本任务讲解高压间歇釜式反应器的操作运行与控制技术，重点阐述高压设备的安全管理、参数调节及应急处理方法。读者将掌握如何确保设备在高压条件下的安全高效运行。

学习目标

知识目标

（1）了解高压间歇釜式反应器的工作原理及特点。

（2）掌握高压操作中的关键控制点和安全措施。

技能目标

（1）能正确操作和维护高压间歇釜式反应器。

（2）能快速识别并处理高压操作中的异常情况。

价值目标

（1）增强高压设备操作的安全意识。

（2）培养对突发事件的应对能力。

High-pressure hydrogenation（氢化）is a common process used in the production of various organic compounds，including pharmaceuticals，fine chemicals，and polymers. The process involves the use of high-pressure hydrogen gas to react with the organic substrate（有机底物）in the presence of a catalyst to achieve hydrogenation. The hydrogenation process is usually carried out in a semi-batch reactor（半间歇反应釜），which is designed to handle the high-pressure conditions and ensure safe operation.

The operation and control of a high-pressure hydrogenation reactors require careful attention to detail and adherence to strict safety protocols（遵守严格的

安全规程）. Before starting the reactor, the operator should ensure that all necessary equipment and materials are in place and that the reactor is properly cleaned and prepared for use. The catalyst should be loaded into the reactor, and the organic substrate should be added to the reactor in a controlled manner.

During the hydrogenation reaction, the temperature, pressure, and hydrogen flow rate should be closely monitored and controlled to ensure optimal reaction conditions. The pressure should be maintained at the desired level using a pressure relief valve（压力释放阀）, and the temperature should be controlled using a cooling system or heating jacket. The hydrogen flow rate should be adjusted to maintain the desired pressure and ensure sufficient hydrogenation of the organic substrate.

After the reaction is complete, the reactor should be carefully depressurized （卸压）to ensure the safe removal of the product. The product should be collected and subjected to a series of purification and separation steps to obtain the desired final product. The reactor should be thoroughly cleaned and prepared for the next batch.

In addition to proper operation and control, safety is a critical consideration in the operation of a high-pressure hydrogenation BR. The operator should be familiar with the safety protocols and procedures and ensure that appropriate safety equipment, such as pressure relief valves and emergency shutdown systems, are in place and functioning properly.

技术理论

生产工艺流程如图 8-1 所示，高压釜经检漏确认密封性良好后，将原料环化物、溶剂醋酸乙酯、催化剂雷尼镍（RNi）加入高压釜中，用氮气置换，然后通入 4～5MPa 的氢气，水浴加热，反应 8～9h 后降温、卸压，含氢化物的上层清液去后处理工序，真空抽滤下层雷尼镍，滤液与上层清液合并，雷尼镍洗涤后回用。

本工艺的 H_2 置换四次后，充入 4～5MPa 的 H_2 压力于釜中，关闭排气阀，用肥皂水进行查漏，检查釜盖上各部接触点是否漏气，包括轴封、排气口等接触部分。检查完毕方可开启搅拌轴封冷却水，然后开启搅拌。

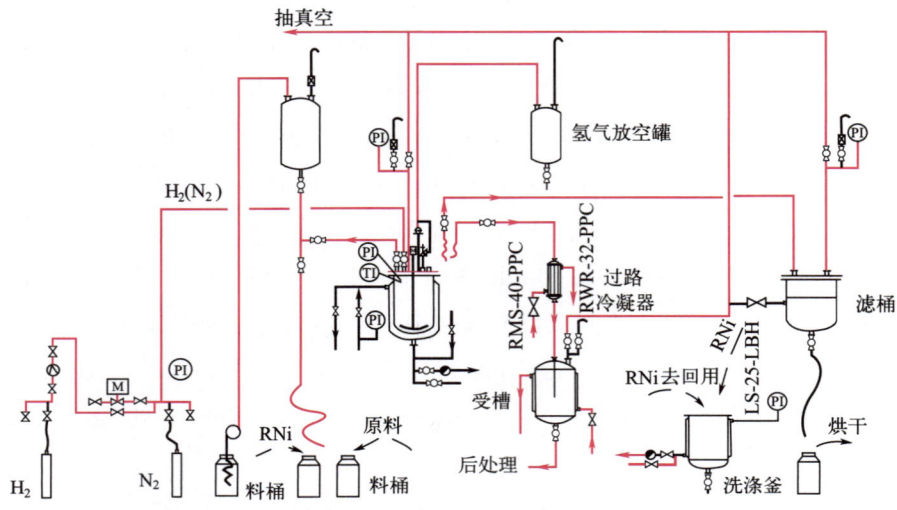

图 8-1　高压氢化反应工艺流程图

关键词详解

检漏，leakage detecting，检查设备是否漏气，高压反应开车之前务必要进行严格的密封性检验。

互动练习

8-1　What is the purpose of high-pressure hydrogenation?

A）To produce inorganic compounds

B）To produce organic compounds

C）To produce polymers

D）To produce metals

8-2　What type of reactor is used for high-pressure hydrogenation?

A）CSTR

B）Semi-batch reactor

C）PFR

D）Packed bed reactor

8-3　What should be done before starting the reactor?

A）Ensure that all necessary equipment and materials are in place

B）Begin loading the catalyst immediately

C）Start adding the organic substrate at a high rate

D）Skip the cleaning and preparation step

8-4　What process parameters should be monitored during hydrogenation reaction?

A）Pressure，temperature，and hydrogen flow rate

B）pH，temperature，and pressure

C）Catalyst concentration，hydrogen flow rate，and pressure

D）Reactor size，temperature，and pressure

8-5　Why is safety a critical consideration in the operation of a high-pressure hydrogenation BR?

A）To ensure efficient production

B）To ensure the product meets quality standards

C）To protect personnel and equipment

D）To reduce costs of production

Operation and Control of CSTRs
连续釜式反应器操作与控制

任务要点

本任务介绍连续釜式反应器的动态行为与控制技术，读者将学习如何调节流量、温度、压力等参数，提高设备运行效率，实现稳定、连续的生产操作。

学习目标

知识目标

（1）掌握连续釜式反应器的动态行为及操作控制方法。

（2）了解常见问题（如积料和温度波动）及其解决方案。

技能目标

（1）能优化连续釜式反应器的运行条件。

（2）能通过数据分析提升生产过程的稳定性。

价值目标

（1）提高连续生产工艺的可靠性。

（2）强化数据驱动的工艺改进能力。

9.1 Stable Operation of CSTRs 连续釜式反应器的稳定操作

CSTRs are commonly used in industrial processes for their ability to maintain a constant reaction rate and ensure product consistency（产品一致性）. Proper operation and control of a CSTR require careful attention to process parameters，particularly heat transfer and stability.

The heat balance equation for a CSTR is essential for calculating the necessary cooling or heating requirements to maintain the reactor at a stable temperature. This equation considers the heat generated or absorbed by the reaction，as well as the heat transferred to or from the surroundings. The heat balance equation can be used to determine the necessary cooling or heating rates and the appropriate cooling or heating equipment.

Stability is also critical for the operation of a CSTR. The stable operation ensures that the reaction rate and product quality are consistent and reliable. Factors that can affect stability include the flow rate and temperature of the reactants，as well as the size of the reactor. It is essential to maintain stable conditions within the reactor by controlling these factors.

The parameters of feed temperature and feed flow rate have a significant impact on the stability of the CSTR. The feed temperature affects the reaction rate and the heat generated by the reaction. A change in feed temperature can cause fluctuations（波动）in the reaction rate，which can lead to instability（不稳定）. Similarly，the feed flow rate affects the reaction rate and the residence time of the reactants within the reactor. A change in the feed flow rate can cause fluctuations in the reaction rate，which can also lead to instability.

技术理论

反应器的可操作性是一个重要问题。影响反应器可操作性的首先是热稳定性。反应器的类型不同，热稳定性的特点也不同。

为讨论热稳定性问题，首先对连续釜式反应器进行热量衡算，通过热量衡算可以得到放热速率与温度的函数关系式，进一步可以得到一条关系曲线，称为反应放热曲线。通过该反应放热曲线即可获得反应的稳定操作点，从而得到热稳定操作的操作参数。

改变反应器的进料温度和进料流量等操作参数，会对热稳定性产生不同的影响。其中低于着火温度和熄火温度分别为进料温度的两个限度。而进料流量的变化也会使得反应"熄火"或者"重燃"。

关键词详解

热稳定性，thermal stability，所谓热稳定性是指反应器本身对热的扰动有无自行恢复平衡的能力。当反应过程的放热或移热因素发生某些变化时，过程的温度等因素将产生一系列的波动，在干扰因素消除后，如果反应过程能恢复到原来的平衡状态，称为热稳定性的，否则称为热不稳定性的。

热量衡算，heat balance，如图 9-1 所示。

连续釜式反应器为一敞开物系，根据热力学第一定律，对于定态操作，其

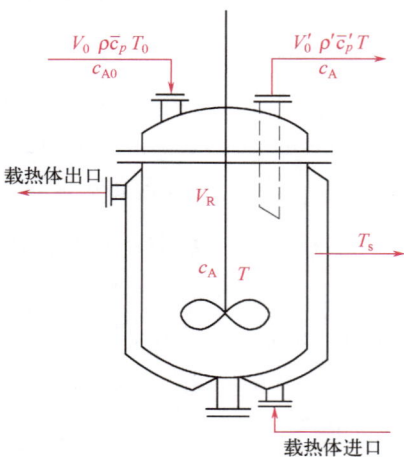

图 9-1 连续釜式反应器热量衡算图

热量衡算式为

$$
\begin{bmatrix} 微元时间内进入 \\ 微元体积的物料 \\ 所带进的热量 \end{bmatrix} = \begin{bmatrix} 微元时间内离开 \\ 微元体积的物料 \\ 带走的热量 \end{bmatrix} - \begin{bmatrix} 微元时间微元 \\ 体积内由于反 \\ 应产生的热量 \end{bmatrix} +
$$

$$
V_0\rho\bar{c}_p(T_0-T_b)\Delta\tau \qquad V'_0\rho'\bar{c}'_p(T-T_b)\Delta\tau \qquad -(-\Delta H_A)(-r_A)V_R\Delta\tau
$$

$$
\begin{bmatrix} 微元时间内微元 \\ 体积传递至环境 \\ 或载热体的热量 \end{bmatrix} + \begin{bmatrix} 微元时间 \\ 微元体积 \\ 内热量的积累 \end{bmatrix}
$$

$$
KA(T-T_s)\Delta\tau \qquad\qquad 0
$$

即：

$$
V_0\bar{c}_p(T_0-T_b)\Delta\tau - V'_0\rho'\bar{c}'_p(T-T_b)\Delta\tau + (-r_A)V_R(-\Delta H_A)\Delta\tau - KA(T-T_s)\Delta\tau = 0
$$

因为
$$
V_0\rho\bar{c}_p \approx V'_0\rho'\bar{c}'_p
$$

故
$$
-V_0\rho\bar{c}_p(T-T_0) + (-r_A)V_R(-\Delta H_A) - KA(T-T_s) = 0
$$

将上式整理，得：

$$
(-r_A)V_R(-\Delta H_A) = V_0\rho\bar{c}_p(T-T_0) + KA(T-T_s) \tag{9-1}
$$

式中 ρ ——反应釜进口物料密度，kg/m^3；

   ρ' ——反应釜出口物料密度，kg/m^3；

   $\bar{c}_p$ —— $T_0 \sim T_b$ 温度范围内物料的平均比热容，$kJ/(kg \cdot K)$；

   $\bar{c}'_p$ —— $T \sim T_b$ 温度范围内物料的平均比热容，$kJ/(kg \cdot K)$；

   V'_0 ——反应釜出口物料体积流量，m^3/h；

   T_b ——基准温度，K。

式（9-1）左端为放热速率 Q_τ

$$Q_\tau = (-r_A)V_R(-\Delta H_A) \tag{9-2}$$

式中，Q_τ 为放热速率，kJ/h。

式（9-2）右端为移热速率 Q_c

$$Q_c = V_0\rho\bar{c}_p(T-T_0) + KA(T-T_s) \tag{9-3}$$

式中，Q_c 为移热速率，kJ/h。

以连续釜式反应器内进行的恒容一级不可逆放热反应为例，其反应速率为 $(-r_A) = kc_A$，则对反应物 A 的物料衡算式（2-10）可改写为

$$c_A = \frac{c_{A0}}{1+k\bar{\tau}} \tag{9-4}$$

联立放热速率式、反应速率式和物料衡算式，可得：

$$Q_\tau = (-\Delta H_A)V_R\frac{kc_{A0}}{1+k\bar{\tau}} = (-\Delta H_A)V_R\frac{F_{A0}k}{V_0(1+k\bar{\tau})} = \frac{k\bar{\tau}F_{A0}(-\Delta H_A)}{1+k\bar{\tau}}$$

$$\tag{9-5}$$

当反应系统及釜内物料转化率确定后，Q_τ 主要取决于反应速率常数 k，即取决于釜内反应物料的温度 T，由阿伦尼乌斯公式可知：

$$k = A_0\exp\left(\frac{-E}{RT}\right)$$

则

$$Q_\tau = \frac{F_{A0}\bar{\tau}(-\Delta H_A)A_0}{\exp\left(\frac{E}{RT}\right)+A_0\bar{\tau}} \tag{9-6}$$

式（9-6）为放热速率 Q_τ 与温度 T 的函数关系式。

反应放热曲线，reaction heat release curve，在 $Q-T$ 坐标图上作一条 S 形曲线，如图 9-2 所示。

在定常态下，单位时间反应放出热量应等于单位时间离开反应釜的物料带出热量与夹套内载热体移出的热量之和，即式（1-95）的移热热速率 Q_c。将式（9-3）在 Q-T 坐标图上可绘制一直线，但因参数值不同，直线有不同的斜率和截距，如图 9-2 中所示的 Q_{c1}、Q_{c2}、Q_{c3} 移热直线。若为绝热过程，式（9-3）右端第二项为零，即

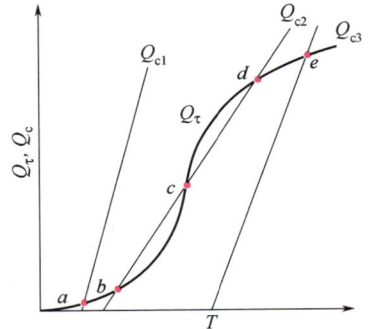

图 9-2 连续釜式反应器的热稳定态示意图

$$Q_c = V_0\rho\bar{c}_p(T-T_0) \tag{9-7}$$

稳定操作点，reactor maintenance，由图 9-2 可以看出，放热曲线与移热直

线 Q_{c2} 的交点有 b、c、d，这三个交点均满足 $Q_\tau = Q_c$，即放热速率与移热速率相等，称为定常状态点。当反应过程中某些因素发生变化或受到干扰，釜温将升高或降低，操作点则偏离定常状态点。对于 b 点（或 d 点），如果温度升高，则 $\dfrac{\mathrm{d}Q_c}{\mathrm{d}T} > \dfrac{\mathrm{d}Q_\tau}{\mathrm{d}T}$，可使釜温下降，恢复到 b 点（或 d 点），反之，釜温下降，则 $\dfrac{\mathrm{d}Q_c}{\mathrm{d}T} < \dfrac{\mathrm{d}Q_\tau}{\mathrm{d}T}$，而使釜温上升恢复到 b 点（或 d 点），故此 b 点和 d 点为热稳定点。而在 c 点操作，外界稍有波动，如釜温升高，$\dfrac{\mathrm{d}Q_c}{\mathrm{d}T} < \dfrac{\mathrm{d}Q_\tau}{\mathrm{d}T}$，将使釜温继续升高至 d 点为止，反之则由于 $\dfrac{\mathrm{d}Q_c}{\mathrm{d}T} > \dfrac{\mathrm{d}Q_\tau}{\mathrm{d}T}$ 使釜温下降到 b 点为止，即在 c 点温度略有升降，系统均不能恢复到原来的热平衡状态。故此 c 点为热不稳定点。综上所述，定常状态稳定操作点必须具备两个条件，即定常状态 $Q_\tau = Q_c$，稳定条件为 $\dfrac{\mathrm{d}Q_c}{\mathrm{d}T} < \dfrac{\mathrm{d}Q_\tau}{\mathrm{d}T}$。

通常，反应釜操作既要维持其操作的稳定性，又希望在适宜的温度下加快反应速率、提高设备生产能力，因此将操作点控制在 d 点为宜。如在 a 点、b 点操作，虽满足热稳定条件，但反应温度偏低，反应速率慢，这是工业上不希望的；而在 e 点操作，反应速率虽然较高，物料的转化率也可提高，但是因为温度较高，副反应增多，收率下降，对热敏性物料也不利。

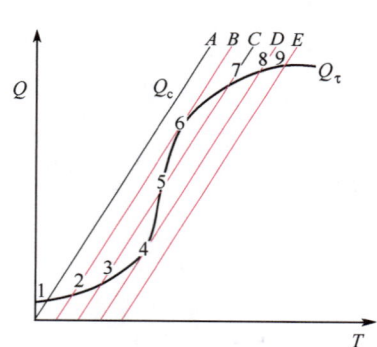

图 9-3 改变进口温度得到不同的操作状态

着火温度，ignition temperature，逐渐改变进料温度 T_0（或冷却介质温度 T_s），则 Q_τ 不变，Q_c 平行移动，如图 9-3 所示。图中相互平行的五条 Q_c 线，表示 5 个不同的进料温度（或冷却介质温度）的移热操作线。当进料温度逐渐提高而使 Q_c 线移至 D 时，它与 Q_τ 线相交于点 4 和点 8，此时只要再略超过 D 线一点，反应器内温度就将骤增至点 8，这时只有一个定常态。根据这一特点，若反应所要求的温度是点 8 处的温度，我们可以使反应器的开车操作沿 D 线迅速达到反应所要求的温度。故在 D 线时的进料温度一般称为着火温度或起燃温度，相应地称点 4 为着火点或起燃点。

熄火温度，quenching temperature，在反应器停车操作时，可逐渐降低 T_0，Q_c 线将沿 D、C、B、A 平行位移，如图 9-3 所示。如果没有较大的温度扰动，反应器内的定态操作温度点 9、8、7、6 变化着。和上述的 D 线情况相似，在降温过程的 B 线，也存在着从点 6 骤降至点 2 的现象。一般称 B 线的温度为熄火温度，点 6 称熄火点。在点 4 和点 6，反应器内出现一种非连续性的温度突变，故在点 4 和点 6 之间，不可能获得稳定操作点。点 4 和点 6 分别是低温操作和高温操作的两个界限。

进料流量，feed flow rate，如图 9-4 所示。当流量从小到大变化时，它们的位置依次为 A、B、C、D、E 等，其操作状态依次变为点 9、8、7、6。当流量稍微超过 D 线所示的量时，定态点立即下跌到点 2，反应被吹"熄"。同样，当流量由高到低变化时，依次得到 1、2、3……各定态点，而在点 4 出现着火现象。操作中，如果由于物料流量过大，而发生熄火现象，可以一面提高进料的温度，一面减小流量，使系统重新点燃。

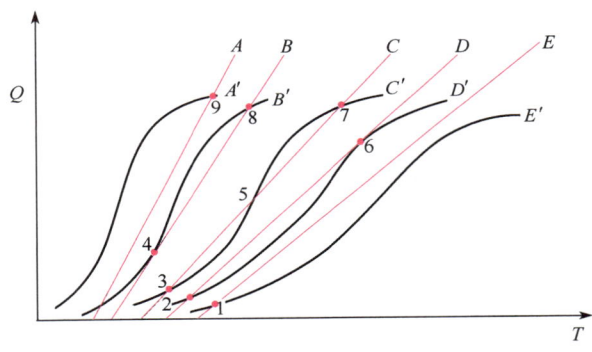

图 9-4　改变进料流量对反应器操作状态的影响

互动练习

9-1　What is the primary benefit of using a CSTR compared to a BR?

A）Greater flexibility in operation

B）Lower capital costs

C）Higher product purity

D）More efficient heat transfer

9-2　What is the heat balance equation for a CSTR?

A）$Q = F_i C_i T_i - F_0 C_0 T_0 - \Delta H_r V_r$

B）$Q = F_0 C_0 T_0 - F_i C_i T_i + \Delta H_r V_r$

C）$Q = F_i C_i T_i - F_0 C_0 T_0 + \Delta H_r V_r$

D）$Q = F_0 C_0 T_0 - F_i C_i T_i - \Delta H_r V_r$

9-3　What is the effect of increasing the feed temperature on the heat stability of a CSTR?

A）Increases heat stability

B）Decreases heat stability

C）Has no effect on heat stability

D）Cannot be determined from the information given

9-4　How does increasing the feed flow rate affect the temperature profile in a CSTR?

A）Increases the temperature of the reactor

B）Decreases the temperature of the reactor

C）Has no effect on the temperature profile

D）Causes fluctuations in the temperature profile

9-5　What is the primary concern when operating a CSTR?

A）Maintaining optimal reaction conditions

B）Ensuring product purity

C）Reducing energy costs

D）Minimizing capital expenditures

9.2　Operation and Control of Polyethylene Continuous Stirred Tank Reactors 聚乙烯连续釜式反应器的操作与控制

Polyethylene is one of the most widely used thermoplastics in the world，and its production relies on CSTR technology. The process involves the polymerization（聚合）of ethylene gas in the presence of a catalyst，which is typically a Ziegler-Natta or metallocene catalyst（茂金属催化剂）.

The CSTR used for polyethylene production typically consists of a reactor vessel equipped with a stirrer，a cooling system，a heating system，and a catalyst injection system（催化剂注入系统）. The reactor is designed to maintain a constant temperature and pressure，and the stirrer is used to ensure efficient mixing of the reactants and to prevent the formation of hot spots（热点）.

The operation and control of the CSTR require careful attention to detail and adherence to strict safety protocols. Before starting the reactor，the operator should ensure that all necessary equipment and materials are in place and that the reactor is properly cleaned and prepared for use. The reactor should be started slowly and monitored carefully to ensure that the temperature, pressure, and agitation are within the desired range.

During the polymerization process, the ethylene gas is fed into the reactor at a controlled rate, and the catalyst is injected into the reactor in a controlled manner. The temperature, pressure, and agitation should be closely monitored and controlled to ensure optimal reaction conditions. Any deviations（偏差）from the desired conditions should be immediately addressed to prevent any adverse effects（负面影响）on the polymerization process.

In the event of an abnormal situation, such as a sudden drop（突然下降）in pressure or temperature, appropriate measures should be taken to prevent any damage to the reactor or the production process. Common abnormal situations in high-density, low-pressure polyethylene production include fouling（结垢）of the reactor, foaming（发泡）of the product, and blockage（堵塞）of the reactor cooling system. These issues can be addressed through various cleaning and maintenance procedures.

技术理论

以图 9-5 所示的生产高密度低压聚乙烯的搅拌釜聚合系统为例，进行连续釜式反应器的操作与控制。通常包括聚乙烯搅拌反应釜的开车、聚合系统的正常操作控制及聚合系统的停车。

乙烯、溶剂己烷以及催化剂、分子量调节剂等连续不断地加入反应器中，在一定的温度、压力条件下进行聚合，聚合热采用夹套及气体外循环、浆液外循环等方式除去，通过调节聚合条件精确控制聚合物的分子量及其分布，反应完成后聚合物浆液靠本身压力出料。

在此稳定运行过程中，要谨防反应温度达到热点，引起飞温，导致反应事故。

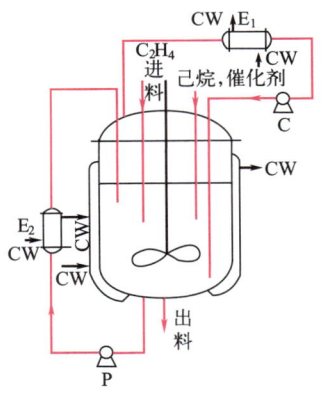

图 9-5　搅拌釜聚合系统示意图
C—循环风机；E—换热器；
P—循环泵；CW—冷却水

关键词详解

聚乙烯搅拌反应釜的开车，thestart-up of polyethylene stirring reactor，首先，通入氮气对聚合系统进行试漏，氮气置换。检查转动设备的润滑情况。投运冷却水、蒸汽、热水、氮气、工厂风、仪表风、润滑油、密封油等系统。投运仪表、电气、安全联锁系统往聚合釜中加入溶剂或液态聚合单体。当釜内液体淹没最低一层搅拌叶后，启动聚合釜搅拌器。继续往釜内加入溶剂或单体，直到达正常料位止。升温使釜温达到正常值。在升温的过程中，当温度达到某一规定值时，向釜内加入催化剂、单体、溶剂、分子量调节剂等，并同时控制聚合温度、压力、聚合釜料位等工艺指标，使之达正常值。

聚合系统的正常操作控制，normal operation and control of the polymeri-zation system，聚乙烯的反应正常操作控制过程包括温度控制、压力控制、液位控制以及聚合浆液浓度控制。

（1）温度控制

聚合温度的控制对于聚合系统操作是最关键的。聚合温度的控制一般有以下三种方法。

① 通过夹套冷却水换热。

② 如图 9-5 所示，循环风机 C、气相换热器 E_1 和聚合釜组成气相外循环系统。通过气相换热器 E_1 能够调节循环气体的温度，并使其中的易冷凝气相冷凝，冷凝液流回聚合釜，从而达到控制聚合温度的目的。

③ 浆液循环泵 P、浆液换热器 E_2 和聚合釜组成浆液外循环系统，通过浆液换热器 E_2 能够调节循环浆液的温度，从而达到控制聚合温度的目的。

（2）压力控制

聚合温度恒定时，在聚合单体为气相时主要通过催化剂的加料量和聚合单体的加料量来控制聚合压力。如聚合单体为液相时，聚合釜压力主要决定单体的蒸气分压，也就是聚合温度。聚合釜气相中，不凝性惰性气体的含量过高是造成聚合釜压力超高的原因之一。此时需放火炬，以降低聚合釜的压力。

（3）液位控制

聚合釜液位应该严格控制。一般聚合釜液位控制在 70% 左右，通过聚合浆液的出料速率来控制。连续聚合时聚合釜必须有自动料位控制系统，以确保液位准确控制。液位控制过低，聚合产率低，液位控制过高，甚至满釜，就会造成聚合浆液进入换热器、风机等设备中，造成事故。

（4）聚合浆液浓度控制

浆液过浓，造成搅拌器电机电流过高，引起超负载跳闸，停转，就会造成釜内聚合物结块，甚至引发飞温，爆聚事故。停搅拌是造成爆聚事故的主要原因之一。如果发生停搅拌事故，应立即加入阻聚剂，并采取其他相应的措施。控制浆液浓度主要通过控制溶剂的加入量和聚合产率来实现的。

聚合系统的停车，shutting down of the polymerization system，该过程需要首先停进催化剂、单体，溶剂继续加入，维持聚合系统继续运行，在聚合反应停止后，停进所有物料，卸料，停搅拌器和其他动设备，用氮气置换，置换合格后交检修。

热点，hot spot，反应的热点是指在化学、物理或其他类型的反应中，产生或积累的热量相对较高的区域或点。这些热点可以是由于反应本身释放的能量，也可以是由于反应速率的增加或其他因素导致的热量聚集。热点是反应过程中的关键部分，因为高温或热量积累可能导致不良的影响，如产生副产物、分解、爆炸或设备损坏。因此，在设计和进行化学反应、实验或工业过程时，识别和管理热点是至关重要的。这可能涉及控制反应速率、调整温度、采取冷却措施、选择合适的反应条件等，以确保反应过程安全、有效和可控。

飞温，temperature-runaway，是指化学反应或过程中由于反应产生的热量变得不受控制，导致温度迅速且无法控制地上升的情况。这可能导致自维持的反应，产生更多热量，形成一个正反馈循环，加速温度上升。飞温可能会带来危险，可能导致设备故障、爆炸或其他危险情况。在本工艺中，一旦发生飞温，应立即停进催化剂、聚合单体，增加溶剂进料量，加大循环冷却水量，紧急放火炬卸压，向后系统排聚合浆液，并适时加入阻聚剂。

互动练习

9-6　What is the purpose of a continuous stirred tank reactor（CSTR）in the production of polyethylene?

A）To add a catalyst to the reaction mixture

B）To ensure the reaction occurs at high pressure

C）To continuously mix the reaction mixture

D）To remove impurities from the product

9-7　What is the role of the initiator in the polyethylene polymerization process?

A）To increase the reaction rate

B）To lower the temperature of the reaction

C）To provide energy for the reaction

D）To remove impurities from the reaction mixture

9-8　How is the temperature of the reaction mixture controlled in a polyethylene CSTR?

A）By adjusting the flow rate of the reaction mixture

B）By adjusting the pressure of the reaction mixture

C）By adding a cooling agent to the reactor

D）By adding a heating agent to the reactor

9-9　What should be done if an abnormal situation occurs during the production of high-density polyethylene?

A）Stop the reaction and drain the reactor

B）Increase the flow rate of the reactants

C）Decrease the temperature of the reaction mixture

D）Add more catalyst to the reactor

9-10　What is the purpose of stopping the CSTR after the completion of the polymerization reaction?

A）To remove impurities from the product

B）To remove the polymer product from the reactor

C）To allow for maintenance of the reactor

D）To prepare for the next polymerization reaction

9.3　Troubleshooting and Maintenance of Tank Reactors
釜式反应器的故障排除与维护

A reactor is a critical piece of equipment in chemical manufacturing，and malfunctions can lead to serious safety hazards and economic losses. Therefore，understanding the common faults and corresponding troubleshooting measures（故障排除措施）is essential for proper reactor operation.

Common problems in reactor operation include material blockage，agitator failure（搅拌故障），leakage，and temperature control issues. For material blockage，the operator should stop the feeding system and clean the blockage. For agitator failure，the operator should stop the system and check the agitator（搅拌），bearings（轴承），and transmission（传动装置）. For leakage，the operator should stop the system and identify the source of the leak. For temperature control issues，the operator should stop the system and check the temperature sensor and controller.

Regular maintenance（日常维护）is crucial for the safe and efficient operation of a reactor. Maintenance should include routine inspection，cleaning，and lubrication of all parts，as well as replacement of worn parts. The operator should also monitor the condition of seals and gaskets and replace them as needed.

Glass-lined reactors are widely used in the chemical industry due to their resistance to chemical corrosion. Normal operation of a glass-lined reactor requires careful handling and maintenance. Operators should avoid excessive temperature changes，impact，and mechanical stress. Cleaning should be carried out regularly using non-abrasive（非研磨性）materials and mild solvents. Proper storage，handling，and transport procedures should be followed to prevent damage to the reactor.

In summary，proper maintenance and troubleshooting are crucial for the safe and efficient operation of a reactor. Operators should be familiar with the common faults and corresponding troubleshooting measures and follow the proper maintenance procedures to ensure the longevity of the reactor.

技术理论

（1）化工生产中搅拌釜式反应器最为常见的故障与处理方法见表 9-1。

表 9-1　釜式反应器常见故障与处理方法

序号	故障现象	故障原因	处理方法
1	壳体损坏（腐蚀、裂纹、透孔）	①受介质腐蚀（点蚀、晶间腐蚀） ②热应力影响产生裂纹或碱脆 ③受损变薄或均匀腐蚀	①用耐蚀材料衬里的壳体需重新修衬或局部补焊 ②焊接后要消除应力，产生裂纹要进行修补 ③超过设计最低的允许厚度需更换本体

续表

序号	故障现象	故障原因	处理方法
2	超温超压	①仪表失灵，控制不严格 ②误操作；原料配比不当；产生剧热反应 ③因传热或搅拌性能不佳，发生副反应 ④进气阀失灵，进气压力过大、压力高	①检查、修复自控系统，严格执行操作规程 ②根据操作法紧急放压，按规定定量、定时投料，严防误操作 ③增加传热面积或清除结垢，改善传热效果；修复搅拌器，提高搅拌效率 ④关总气阀，切断气源，修理阀门
3	密封泄漏	填料密封 ①搅拌轴在填料处磨损或腐蚀，造成间隙过大 ②油环位置不当或油路堵塞，不能形成油封 ③压盖没压紧，填料质量差或使用过久 ④填料箱腐蚀（机械密封） ⑤动静环端面变形、碰伤 ⑥端面比压过大，摩擦产生热导致变形 ⑦密封圈选材不对，压紧力不够，或V形密封圈装反，失去密封性 ⑧轴线与静环端面垂直度误差过大 ⑨操作压力、温度不稳，硬颗粒进入摩擦副 ⑩轴窜量超过指标 ⑪镶装或粘接动、静环的镶缝泄漏	①更换或修补搅拌轴，并在机床上加工，保证表面粗糙度 ②调整油环位置，清洗油路 ③压紧填料或更换填料 ④修补或更换 ⑤更换摩擦副或重新研磨 ⑥调整比压要合适，加强冷却系统，及时带走热量 ⑦密封圈选材、安装要合理，要有足够的压紧力 ⑧停车，重新找正，保证垂直度误差小于0.5mm ⑨严格控制工艺指标，颗粒及结晶物不能进入摩擦副 ⑩调整、检修，使轴的窜量达到标准 ⑪改进安装工艺，或过盈量要适当，或粘接剂要好用，粘接牢固
4	釜内有异常杂声	①搅拌器摩擦釜内附件（蛇管、温度计管等）或刮壁 ②搅拌器松脱 ③衬里鼓包，与搅拌器撞击 ④搅拌器轴弯曲或轴承损坏	①停车检修找正，使搅拌器与附件有一定间距 ②停车检查，紧固螺栓 ③修鼓包或更换衬里 ④检修或更换轴及轴承
5	搪瓷搅拌器脱落	①被介质腐蚀断裂 ②电动机旋转方向相反	①更换搪瓷轴或用玻璃修补 ②停车改变转向

续表

序号	故障现象	故障原因	处理方法
6	搪瓷釜法兰漏气	①法兰瓷面损坏 ②选择垫圈材质不合理,安装接头不正确,空位,错移 ③卡子松动或数量不足	①修补、涂防腐漆或树脂 ②根据工艺要求选择垫圈材料,垫圈接口要搭拢,位置要均匀 ③按设计要求有足够数量的卡子,并要紧固
7	瓷面产生鳞爆及微孔	①夹套或搅拌轴管内进入酸性杂质,产生氢脆现象 ②瓷层不致密,有微孔隐患	①用碳酸钠中和后,用水冲净或修补,腐蚀严重的需更换 ②微孔数量少的可修补,严重的更换
8	电动机电流超过额定值	①轴承损坏 ②釜内温度低,物料黏稠 ③主轴转数较快 ④搅拌器直径过大	①更换轴承 ②按操作规程调整温度,物料黏度不能过大 ③控制主轴转数在一定的范围内 ④适当调整检修

（2）釜式反应器维护要点

① 反应釜在运行中，严格执行操作规程，禁止超温、超压。

② 按工艺指标控制夹套（或蛇管）及反应器的温度。

③ 避免温差应力与内压应力叠加，使设备产生应变。

④ 要严格控制配料比，防止剧烈反应。

⑤ 要注意反应釜有无异常振动和声响，如发现故障，应检查修理并及时消除。

（3）搪玻璃反应釜正常操作要点

① 加料要严防金属硬物掉入设备内，运转时要防止设备振动，检修时按化工厂搪玻璃反应釜维护检修规程（HGJ1008—79）执行。

② 尽量避免冷罐加热料和热罐加冷料，严防温度骤冷骤热，搪玻璃耐温剧变小于120℃。

③ 尽量避免酸碱液介质交替使用，否则将会使搪玻璃表面失去光泽而腐蚀。

④ 严防夹套内进入酸液（如果清洗夹套一定要用酸液时，不能用 pH＜2 的酸液），酸液进入夹套会产生氢效应，引起搪玻璃表面像鱼鳞片一样大面积

脱落。一般清洗夹套可用2%的次氯酸钠溶液，最后用水清洗夹套。

⑤ 出料釜底堵塞时，可用非金属棒轻轻疏通，禁止用金属工具铲打。对粘在罐内表面上的反应物料要及时清洗，不宜用金属工具，以防损坏搪玻璃衬里。

关键词详解

故障排除措施，*troubleshooting measures*，反应设备的故障排除措施是指在化学、物理或工艺反应中出现问题或异常情况时，采取的步骤和方法来确定问题原因并修复设备，以使反应能够恢复正常运行。简要的故障排除措施的介绍如下。

• 问题诊断：首先，需要仔细检查设备和反应过程，确定具体的问题，例如温度异常、压力波动、搅拌异常等。

• 停机操作：在发现问题后，可能需要立即停止反应设备，避免进一步的损坏或危险。

• 检查设备：对设备的各个组件、管道、阀门等进行检查，确认是否有损坏、松动或其他异常。

• 数据分析：分析实时监测的数据，例如温度、压力、流量等，以找出可能的问题源。

• 操作参数调整：如果问题是由于操作参数不当引起的，可以尝试调整参数，例如温度、压力、搅拌速度等。

• 部件更换：如果发现有损坏的部件，需要及时更换，确保设备的正常运行。

• 清洁维护：反应设备可能因为杂质积聚或污染而出现问题，进行适当的清洁和维护可以解决一些故障。

• 紧急修复：对于需要立即修复的问题，可能需要进行临时的紧急修复，以确保设备的安全和继续运行。

• 专业支持：如果问题复杂或无法解决，可能需要寻求专业工程师或技术人员的帮助，进行更深入的分析和修复。

• 故障分析报告：在故障排除完成后，应该进行故障分析，总结问题原因和解决方法，以便以后避免类似问题的发生。

互动练习

9-11　Why is understanding common faults and corresponding trouble-shooting measures important for reactor operation?

A）To increase profits

B）To avoid safety hazards and economic losses

C）To improve equipment aesthetics

D）To save time on maintenance

9-12　Which of the following is NOT a common problem in reactor operation?

A）Material blockage

B）Agitator failure

C）Temperature control success

D）Leakage

9-13　What should the operator do in case of material blockage?

A）Stop the feeding system and clean the blockage

B）Increase the feeding rate to clear the blockage

C）Continue operation and hope the blockage clears on its own

D）Ignore the blockage and continue operation

9-14　What should be done to maintain the proper operation of a glass-lined reactor?

A）Expose it to excessive temperature changes

B）Use abrasive cleaning materials

C）Avoid mechanical stress

D）Store and transport it carelessly

9-15　What should be included in regular maintenance of a reactor?

A）Only replacement of worn parts

B）Routine inspection，cleaning，and lubrication of all parts

C）Replacement of seals and gaskets as needed

D）None of the above

任务10

Operation and Control of PFRs Reactors
连续管式反应器操作与控制

任务要点

本任务涵盖连续管式反应器的操作与控制技术，读者将学习流体力学与热传导的基本原理，并通过实际案例了解如何优化运行参数，提高工业生产效率。

学习目标

知识目标

(1) 了解连续管式反应器的流体力学及热传导特性。

(2) 掌握连续管式反应器操作中的关键参数控制方法。

技能目标

(1) 能优化连续管式反应器的工艺条件以提高产能和质量。

(2) 能诊断并解决管式反应器运行中常见的问题。

价值目标

(1) 强化安全生产与节能环保的意识。

(2) 培养设备操作中的创新性思维。

Epoxy ethane（EO）（环氧乙烷）is widely used in the production of various chemicals，and one of its main products is ethylene glycol（EG）（乙二醇），which is produced by reacting EO with water in a PFR. The process involves the reaction of EO with water at high temperature and pressure in the presence of a catalyst to produce EG. The process is exothermic，and the heat generated must be carefully controlled to avoid overheating and thermal runaway（热失控）.

Before starting the reactor，the operator must perform a thorough inspection of the equipment and ensure that all necessary materials are in place. Once the reactor is ready，the operator can start the process by introducing the reactants into the reactor at the desired flow rates. The reactor should be heated to the desired temperature，and the pressure should be maintained at the desired level using a pressure control system.

During the reaction，the temperature and pressure must be closely monitored and controlled to ensure safe and efficient operation. If any abnormalities are detected，the operator should take appropriate measures to address them，such as adjusting the temperature or pressure，or stopping the process.

After the reaction is complete，the operator can stop the flow of reactants and products and initiate the cooling process. The reactor should be cooled down（冷却下来）to a safe temperature before it can be opened for cleaning and maintenance.

Common abnormalities in the operation of the PFR include fouling，coking（焦化），and corrosion（腐蚀）. These issues can be caused by a variety of factors，including improper operation，poor maintenance，or poor quality of the reactants or catalyst. To avoid these issues，the operator should follow the proper maintenance procedures and perform regular inspections and cleaning of the reactor and associated equipment（相关设备）.

技术理论

环氧乙烷与水反应流程如图 10-1 所示，精制塔塔底物料在流量控制下同循环水排放物流以 1∶22 的摩尔比混合，混合后通过在线混合器进入乙二醇反应器。反应为放热反应，反应温度为 200℃ 时，每生成 1mol 乙二醇放出热量为 $8.315×10^4$ J。来自循环水排放浓缩器的水，是在同精制塔塔底物料的流量比控制下进入乙二醇反应器上游的在线混合器的。混合物流通过乙二醇反应器，在此反应，形成乙二醇。反应器的出口压力是通过维持背压来控制的。从乙二醇反应器流出的乙二醇-水物流进入干燥塔。

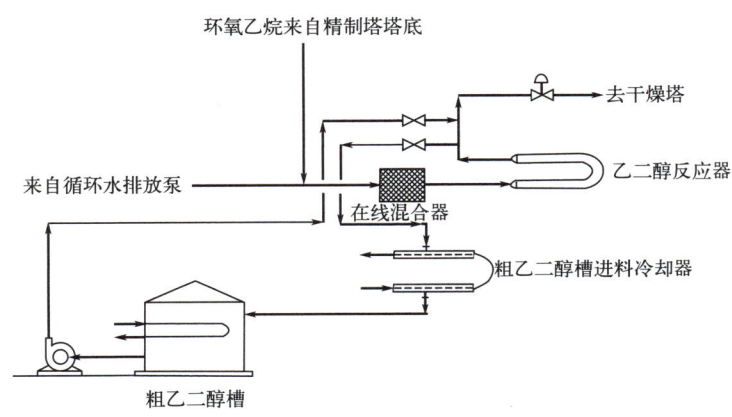

图 10-1　乙二醇生产工艺流程图

关键词详解

进料组成，feed composition，是指反应物料中的各组成配比，本工艺的乙二醇反应器进料组成是通过控制循环水排放到混合器的流量和精制塔内环氧乙烷排放到混合器的流量的比例来实现的。

反应器维护，reactor maintenance，是指对反应设备进行的日常及定期的管理、维护、保养等。相对于釜式反应器，管式反应器一般没有搅拌器一类转动部件，故具有密封可靠、振动小的特点，反应器维护比较简便。

互动练习

10-1　What is the purpose of the PFR in the production of ethylene glycol?

A）To mix the reactants

B）To increase the reaction time

C）To cool down the reactants

D）To heat up the reactants

10-2　What are some common abnormal phenomena in the production of ethylene glycol?

A）High reaction temperature and low pressure

B）Low reaction temperature and high pressure

C）High reaction temperature and high pressure

D）Low reaction temperature and low pressure

10-3　What should be done before starting the reactor for the production of ethylene glycol?

A）Load the reactants into the reactor

B）Check and prepare the reactor

C）Collect and purify the product

D）Adjust the reaction temperature and pressure

10-4　What is the purpose of the heat exchanger in the production of ethylene glycol?

A）To increase the reaction rate

B）To mix the reactants

C）To remove heat from the reaction

D）To increase the pressure of the reaction

10-5　　What are some common maintenance tasks for the PFR used in the production of ethylene glycol?

A）Cleaning the reactor and replacing the catalyst

B）Adjusting the reaction temperature and pressure

C）Adding more reactants to the reactor

D）Increasing the flow rate of the reactants

项目二

Selection，Design，Operation and Control of Gas-Solid Phase Reactors

气固相反应器选择、设计、操作与控制

项目二

Selection，Design，Operation and Control of Gas-Solid Phase Reactors
气固相反应器选择、设计、操作与控制

任务11

Selection of Gas-Solid Phase Reactors
气固相反应器选择

任务要点

本任务主要讲解气固相反应器的分类及适用场景，重点分析固定床和流化床反应器的特性和使用条件。读者将学习如何根据具体的化工工艺需求选择合适的气固相反应器，并掌握设备选型对生产效率和产品质量的影响。

学习目标

知识目标

（1）了解气固相反应器的类型及其适用场景。

（2）掌握固定床和流化床反应器的基本概念。

技能目标

（1）能根据生产要求选择适合的气固相反应器。

（2）能评价不同反应器在操作中的优缺点。

价值目标

（1）在设备选择中平衡生产成本与效果。

（2）强调绿色化工生产的重要性。

11.1 Characteristics and Structures of Fixed Bed Reactors
固定床反应器特点与结构

Fixed bed reactors（固定床反应器）are commonly used in chemical processes that involve catalytic reactions. These reactors have a stationary（固定的）bed of solid catalysts through which the reactants flow. Fixed bed reactors offer several advantages，including high reaction rates，and reduced risk of catalyst attrition（催化剂损耗）. However，they also have some disadvantages，such as hard catalyst replacement（催化剂更换困难），pressure drop（压降）and temperature gradients（温度梯度）.

There are two main types of fixed bed reactors based on their structures：adiabatic（绝热式）and heat exchange（换热式）. Adiabatic fixed bed reactors are typically used for reactions with insignificant thermal effects（不明显热效应），while non-adiabatic reactors are used for reactions with significant thermal effect（明显热效应）. In reaction zone，adiabatic reactors do not have any heat exchange mechanisms，while non-adiabatic reactors use heat exchangers to maintain the reaction temperature.

The structure of a fixed bed reactor typically consists of a cylindrical or rectangular vessel（圆柱形或矩形容器）filled with catalyst particles. The reactants are fed from the top of the reactor and flow through the catalyst bed，while the products are collected at the bottom. The reactor may also have a preheater（预热器）to raise the reactant temperature before entering the catalyst bed.

The catalyst bed is typically supported by a distributor plate（分布板）that ensures uniform flow distribution（均一的流体分布）and prevents catalyst attrition. The bed may also have a retaining screen at the bottom to prevent the catalyst from escaping.

In summary，fixed bed reactors are widely used in chemical processes that involve catalytic reactions. They offer several advantages but also have some drawbacks. The reactor's structure depends on the type of reaction and may include a preheater，distributor plate，and retaining screen（固定滤网）. Adiabatic and non-adiabatic reactors differ in their heat exchange mechanisms.

技术理论

流体通过不动的固体物料形成的床层面进行反应的设备都称为固定床反应器，其中尤以利用气态的反应物料，通过由固体催化剂构成的床层进行反应的气固相催化反应器在化工生产中应用最为广泛。这类反应器的特点是充填在设备内的固体颗粒固定不动，有别于固体物料在设备内发生运动的移动床和流化床，又称填充床反应器。固定床反应器广泛用于气固相反应和液固相反应过程。

随着化工生产技术的进步，已出现多种固定床反应器的结构类型，以适应不同的传热要求和传热方式，主要分为绝热式和换热式两类。

绝热式固定床反应器结构简单，催化剂均匀装填于床层内，反应器绝热措施良好，无热量损失且与外界无热量交换。绝热式反应器分为单段绝热式固定床反应器和多段绝热式固定床反应器。

换热式固定床反应器以列管式为多，通常管内装催化剂，管间走载热体，当反应热效应较大时，为了维持适宜的温度条件，必须利用换热介质移走或供给热量。按换热介质不同，可分为对外换热式固定床反应器和自热式固定床反应器。

关键词详解

固定床反应器，fixed bed reactor，指在反应器内装填颗粒状固体催化剂或固体反应物，形成一定高度的堆积床层，气体或液体物料通过颗粒间隙流过静止固定床层的同时，实现非均相反应过程。

固体催化剂，solid catalyst，在化学反应里能改变反应物化学反应速率（提高或降低）而不改变化学平衡，且本身的质量和化学性质在化学反应前后都没有发生改变的固体物质叫固体催化剂。固体催化剂是现代催化技术发展的一个方向，其中最有代表性的当属固体酸、固体碱的工业化应用。

绝热式固定床反应器，adiabatic fixed bed reactor，如图11-1所示，反应器外壳包裹绝热保温层，使催化剂床层与外界没有热量交换。中空圆筒的底部放置搁板，上面堆放固体催化剂。气体从上而下通过催化剂床层。结构简单，床层横截面温度均匀。单位体积内催化剂量大，即生产能力大。但只适用于热效应不大的反应。图11-2所示为多段绝热式固定床反应器。

换热式固定床反应器，heat exchange fixed bed reactor，此设备如同列管式换热器，又称为列管式固定床反应器。反应器由多根反应管并联构成，管径一般为25～30mm，管数可达万根以上。管内装催化剂，传热介质流经管间进行加热或冷却。图11-3所示为换热式固定床反应器最常用的类型——列管式固定床反应器。

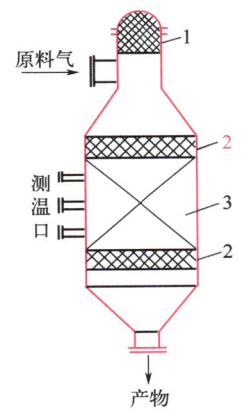

动画

绝热式固定床反应器

图11-1 绝热式固定床反应器
1—矿渣棉；2—瓷环；3—催化剂

自热式固定床反应器，self-heating fixed bed reactor，采用反应放出的热量来预热新鲜的进料，达到热量自给和平衡，其设备紧凑，可用于高压反应体系。但其结构较复杂，操作弹性较小，启动反应时常用电加热。图11-4径向固定床催化反应器中，不仅降低了进出口气体压降，还将进气与出气在反应器下半段进行了换热。

动画

中间换热式
固定床反应器
冷激式固定
床反应器

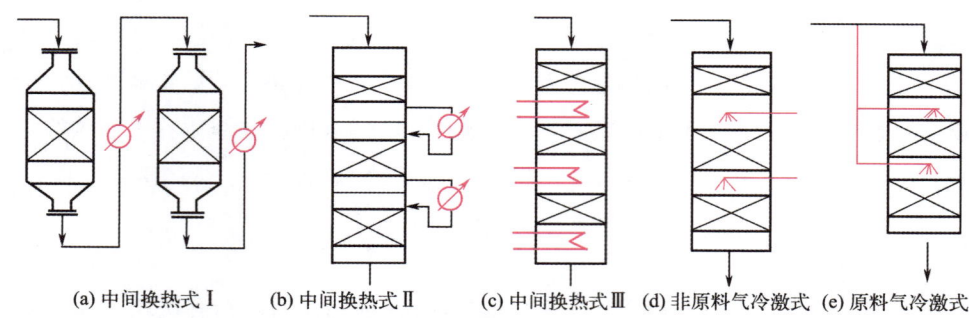

(a) 中间换热式Ⅰ   (b) 中间换热式Ⅱ   (c) 中间换热式Ⅲ   (d) 非原料气冷激式   (e) 原料气冷激式

图 11-2   多段绝热式固定床反应器

动画

列管式固定
床反应器
径向固定床
催化反应器

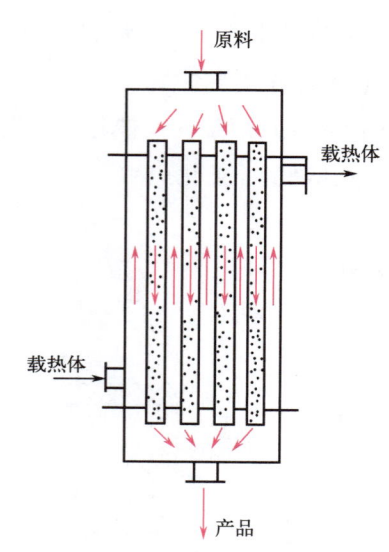

图 11-3   列管式固定床反应器

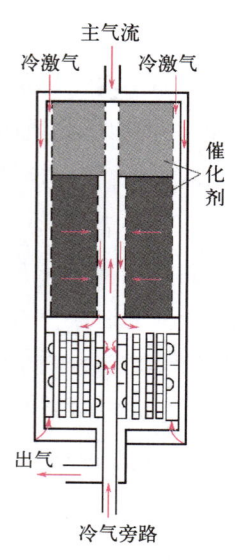

图 11-4   径向固定床催化反应器

动画

列管式固定
床反应器的
温度分布

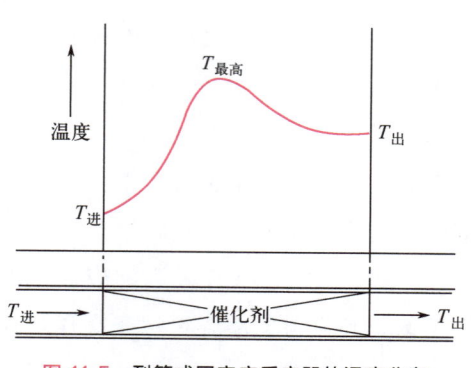

图 11-5   列管式固定床反应器的温度分布

热点温度，hot-spot tempera-ture，对于强放热的反应如氧化反应，径向和轴向都有温差。如催化剂的导热性能良好，而气体流速又较快，则径向温差可较小。轴向的温度分布主要决定于沿轴向各点的放热速率和管外载热体的移热速率。一般沿轴向温度分布都有一最高温度，称为热点，如图 11-5 所示。

在热点以前放热速率大于移热速率，因此出现轴向床层温度升高，热点以后恰恰相反，故沿床层温度逐渐降低。控制热点温度是使反应能顺利进行的关键。热点温度过高，使反应选择性降低，催化剂变劣，甚至使反应失去稳定性而产生飞温。热点出现的位置及高度与反应条件的控制、传热和催化剂的活性有关。随着催化剂的逐渐老化，热点温度逐渐下移，其高度也逐渐降低。

为了降低热点温度，减少轴向温差，使沿轴向大部分催化剂床层能在适宜的温度范围内操作，工业生产上所采取的措施有：①在原料气中带入微量抑制剂，使催化剂部分毒化；②在原料气入口处附近的反应管上层放置一定高度为惰性载体稀释的催化剂，或放置一定高度已部分老化的催化剂，这两项措施目的是降低入口处附近的反应速率，以降低放热速率，使与移热速率尽可能平衡；③采用分段冷却法，改变移热速率，使与放热速率尽可能平衡等。

互动练习

11-1　What is the purpose of using fixed bed reactors?

A）To reduce the reaction rate

B）To increase the risk of catalyst attrition

C）To have a stationary bed of solid catalysts

D）To avoid the use of heat exchangers

11-2　Which of the following is an advantage of fixed bed reactors?

A）High pressure drops

B）Low reaction rates

C）Reduced risk of catalyst attrition

D）Complex catalyst replacement

11-3　Which type of fixed bed reactor is typically used for exothermic reactions?

A）Adiabatic

B）Heat exchange

C）Both

D）None of the above

11-4　What is the function of a distributor plate in a fixed bed reactor?

A）To ensure uniform flow distribution

B）To prevent catalyst attrition

C）To retain the catalyst

D）To maintain the reaction temperature

11-5　In a fixed-bed reactor，what does the hot spot temperature typically refer to?

A）The temperature at the reactor inlet

B）The temperature at the reactor outlet

C）The point with the highest temperature within the reactor bed

D）The temperature at the reactor wall

11.2　Characteristics and Structures of Fluidized Bed Reactors 流化床反应器特点与结构

Fluidized bed reactors（流化床反应器）are widely used in chemical processes that involve gas-solid reactions. These reactors have a bed of solid particles that are suspended in a gas stream，which causes the particles to behave like a fluid. Fluidized bed reactors offer several advantages，including high heat and mass transfer rates（更高的传热传质速率），reduced risk of hot spots（热点的风险降低），and better catalyst utilization（催化剂使用便捷）. However，they also have some disadvantages，such as the need for careful control of the gas flow（需要谨慎控制气体流速），high capital and operating costs（投资费用和操作费用高），and the risk of particle attrition（颗粒磨损）.

There are several types of fluidized bed reactors based on their structures. The main components of a fluidized bed reactor include the main body（主体），gas distribution device（气体分布装置），internal components（内部组件），heat exchange device，and gas-solid separation device（气固分离装置）. The main body is typically cylindrical or rectangular and is filled with solid particles that are fluidized by the gas stream. The gas distribution device is used to distribute the gas evenly throughout the bed and prevent channeling（防止沟流）. Internal components such as baffles and grids are used to enhance mixing and reduce dead zones（防止产生死区）. Heat exchange devices such as coils or plates are used to maintain the reaction temperature. Finally，gas-solid separation devices such as cyclones（旋风分离器）or filters are used to separate the product gases from the solid particles.

In summary，fluidized bed reactors are widely used in chemical processes due to their high heat and mass transfer rates，better catalyst utilization，and reduced risk of hot spots. The reactor's structure includes several components such as the main body，gas distribution device，internal components，heat exchange device，and gas-solid separation device. However，they also have some drawbacks such as high capital and operating costs and the risk of particle attrition.

技术理论

流化床反应器是利用流体通过颗粒状固体层而使固体颗粒处于悬浮运动状态的反应器。流化床反应器是工业上较为广泛应用的一类反应器，适用于催化或非催化的气固、液固和气液固反应。

流化床反应器具有较大的接触面积，能够实现高效的物质传递和热交换，提高了反应效率。流化床反应器的颗粒床具有很好的温度均匀性，可有效控制反应温度，降低产物的副反应。流化床反应器内的颗粒床可以有效地吸收和释放热量，使得反应器具有良好的热稳定性，并能够适应多种反应条件。此外，流化床反应器还具有操作灵活性高、可连续生产、易于扩大规模等特点。

流化床反应器的结构主要包括：反应器本体、气体分布装置、内部构件、换热装置和气固分离装置等。

关键词详解

流化床反应器，fluidized bed reactor，如图 11-6 所示，是指气体在由固体物料或催化剂构成的沸腾床层内进行化学反应的设备。又称"沸腾床反应器"。气体在一定的流速范围内，将堆成一定厚度（床层）的催化剂或物料的固体细粒强烈搅动，使之像沸腾的液体一样并具有液体的一些特性，如对器壁有流体压力的作用、能溢流和具有黏度等，此种操作状况称为"流化床"。反应器上部有扩大段，内装旋风分离器，用以回收被气体带走的催化剂；底部设置原料进口管和气体分布板；中部为反应段，装有冷却水管和导向挡板，用以控制反应温度和改善气固接触条件。

连续生产，continuous production，连续操作设备利用率高，产品质量稳定，易于自动控制，适用于大规模生产。

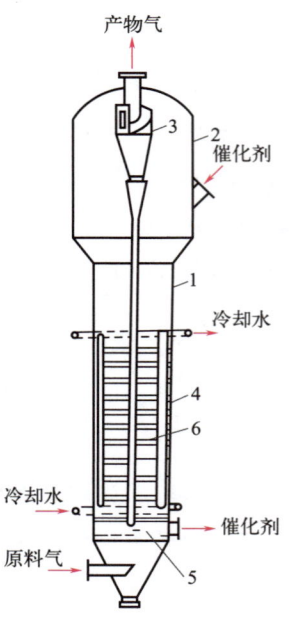

图 11-6　流化床反应器

1—壳体；2—扩大段；3—旋风分离器；4—换热管；5—气体分布器；6—内部构件

流化床反应器

气体分布装置，gas distribution device，气体分布装置包括设置在锥底的气体预分布器和气体分布板两部分。其作用是使气体均匀分布，以形成良好的初始流化条件，同时支承固体催化剂颗粒。

内部构件，internals，内部构件一般设置在浓相段，主要用来破碎气体在床层中产生的大气泡，增大气固相间的接触机会；减少返混，从而增加反应速率和提高转化率。内部构件包括挡网、挡板和填充物等。在气流速率较低、催化反应对于产品要求不高时，可以不设置内部构件。

换热装置，heat exchange device，换热装置的作用是用来取出或供给反应所需要的热量。由于流化床反应器的传热速率远远高于固定床，因此同样反应所需的换热装置要比固定床中的换热装置小得多。根据需要分为外夹套换热器和内管换热器，也可采用电感加热。图 11-7 为流化床常用的内部换热器。

气固分离装置，gas-solid separator，由于流化床内的固体颗粒不断地运动，引起粒子间及粒子与器壁间的碰撞而磨损，使上升气流中带有细粒和粉尘。气固分离装置用来回收这部分细粒，使其返回床层，并避免带出粉尘影响产品纯度。常用的气固分离装置有旋风分离器和过滤管。

旋风分离器是一种靠离心作用把固体颗粒和气体分开的装置，结构如图11-8 所示。含有催化剂颗粒的气体由进气管沿切线方向进入旋风分离器内，在

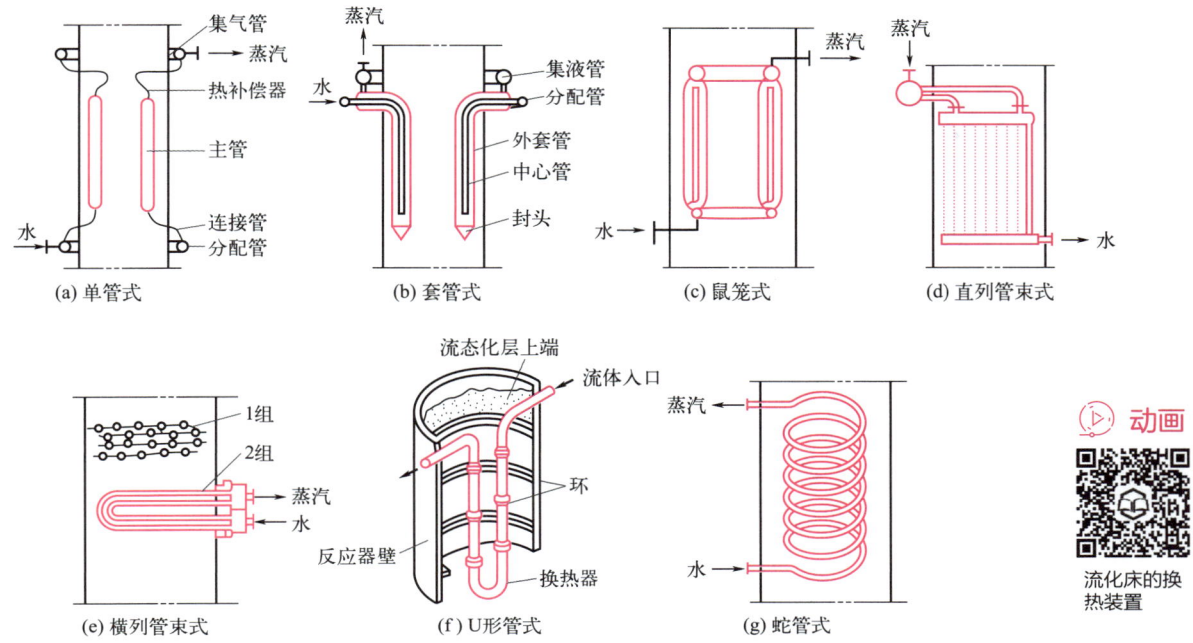

图 11-7　流化床常用的内部换热器

旋风分离器内作回旋运动而产生离心力，催化剂颗粒在离心力的作用下被抛向器壁，与器壁相撞后，借重力沉降到锥底，而气体则由上部排气管排出。为了加强分离效果，有些流化床反应器在设备中把三个旋风分离器串联使用，催化剂按大小不同的颗粒先后沉降至各级分离器锥底。

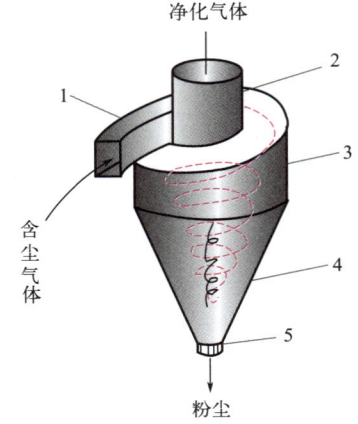

图 11-8　旋风分离器结构示意图

1—矩形进口管；2—螺旋状进口管；3—筒体；4—锥体；5—灰斗

动画

流化床的换热装置

动画

旋风分离器

互动练习

11-6   What is the main characteristic of a fluidized bed reactor?

A）It has a fixed bed of solid catalysts

B）It uses a fluid to suspend the catalyst particles

C）It has apreheater to raise the reactant temperature

D）It uses heat exchangers to maintain the reaction temperature

11-7   What is the advantage of using a fluidized bed reactor?

A）It has a reduced risk of catalyst attrition

B）It has a lower pressure drop than fixed bed reactors

C）It offers high reaction rates

D）It does not require a retaining screen to prevent catalyst escape

11-8   What is the function of the distributor plate in a fluidized bed reactor?

A）It ensures uniform flow distribution and prevents catalyst attrition

B）It separates the reactants from the products

C）It raises the temperature of the reactants

D）It prevents the catalyst from escaping

11-9   What is the purpose of the gas distribution device in a fluidized bed reactor?

A）It distributes the gas evenly throughout the reactor

B）It separates the gas from the solid catalysts

C）It preheats the gas before it enters the reactor

D）It maintains the reaction temperature

11-10   Through which action does a cyclone separator mainly allow particles to settle and thus achieve separation?

A）Thermal effect

B）Inertial centrifugal force effect

C）Chemical reaction effect

D）Gravity effect

11.3 Selection of Gas-Solid Catalytic Reactors 气固相催化反应器选择

Gas-solid catalytic reactors（气固相催化反应器）are widely used in the chemical industry for carrying out catalytic reactions between gases and solids. The selection of a suitable reactor for a particular reaction is a critical task，and several factors must be considered before making a choice.

Firstly，the reaction characteristics（反应特征），such as reaction kinetics，selectivity，and yield，must be considered to determine the reaction rate and the products' purity.

Secondly，the reaction heat must be taken into account to ensure the reactor's thermal stability and to prevent thermal runaway（热失控）. The thermal stability can be maintained by controlling the reaction temperature through proper heat transfer mechanisms.

Thirdly，the process requirements（工艺需求），such as pressure，reactant flow rate，and contact time，must be considered to ensure that the reactor can meet the process requirements.

Fourthly，the reactor's design must consider its features，such as heat and mass transfer rates，catalyst activity and selectivity，and pressure drop across the reactor.

Finally，the catalyst's properties，such as specific surface area（表面积），pore size distribution（孔径分布），and activity，must be considered to ensure its effectiveness in the reaction.

In summary，selecting a gas-solid catalytic reactor requires a comprehensive understanding of the reaction's characteristics，thermal stability，process requirements，reactor design，and catalyst properties. Only by considering these factors can a suitable reactor be chosen for a specific catalytic reaction.

技术理论

气固相催化反应器的选择一般可从反应特点、反应热、工艺要求、反应器特点、催化剂性能等方面综合考量。表 11-1 所示为气固相催化反应器选择举例。

表 11-1　气固相催化反应器选择举例

类型	适用的反应	应用特点	应用举例
固定床	气固（催化或非催化）相	返混小，高转化率时催化剂用量少，催化剂不易磨损，但传热控温不易，催化剂装卸麻烦	乙苯脱氢制苯乙烯，乙炔法制氯乙烯，合成氨，乙烯法制醋酸乙烯等
流化床	气固（催化或非催化）相	传热好，温度均匀，易控制，催化剂有效系数大，粒子输送容易，但磨耗大，床内返混大，对高转化率不利，操作条件限制较大	萘氧化制苯酐，石油催化裂化，乙烯氧氯化制二氯乙烷等
移动床	气固（催化、非催化）相	固体返混小，固气比可变性大，但粒子传送较易，床内温差大，调节困难	石油催化裂化，矿物的焙烧或冶炼

关键词详解

气固相催化反应器，gas-solid catalytic reactor，气固催化反应器是近代化学工业上最普遍采用的反应器之一。涉及两个相，即气相与固相，因而它是非均相反应器。

反应热，reaction heats，反应热是指当一个化学反应在恒压以及不做非膨胀功的情况下发生后，若使生成物的温度回到反应物的起始温度，这时体系所放出或吸收的热量称为反应热。也就是说，反应热通常是指：体系在等温、等压过程中发生化学的变化时所放出或吸收的热量。化学反应热有多种形式，如：生成热、燃烧热、中和热等。

移动床，moving bed reactor，是一种用以实现气固相反应过程或液固相反应过程的反应器。在反应器顶部连续加入颗粒状或块状固体反应物或催化剂，随着反应的进行，固体物料逐渐下移，最后自底部连续卸出。流体则自下而上（或自上而下）通过固体床层，以进行反应。由于固体颗粒之间基本上没有相对运动，但却有固体颗粒层的下移运动。因此，也可将其看成是一种移动的固定床反应器。

互动练习

11-11　What is the main consideration when selecting a gas-solid catalytic reactor?

A）The type of reaction involved

B）The size of the reactor

C）The location of the reactor

D）The cost of the reactor

11-12　Why is the heat generated in gas-solid catalytic reactions a concern?

A）It can lead to a decrease in the reaction rate

B）It can damage the reactor

C）It can cause a temperature gradient within the reactor

D）It has no effect on the reaction

11-13　What are the typical process requirements for gas-solid catalytic reactions?

A）High pressure and high temperature

B）Low pressure and low temperature

C）High pressure and low temperature

D）Low pressure and high temperature

11-14　What are the main characteristics of gas-solid catalytic reactors?

A）They have a stationary bed of catalysts

B）They are used for endothermic reactions

C）They have a fluidized bed of catalysts

D）They are used for exothermic reactions

11-15　What is an important consideration for the performance of the catalyst in a gas-solid catalytic reaction?

A）The size of the catalyst particles

B）The color of the catalyst particles

C）The shape of the catalyst particles

D）The chemical composition of the catalyst particles

任务12

Design of Fixed Bed Reactors
固定床反应器设计

任务要点

本任务聚焦固定床反应器的设计原理，介绍其关键设计参数（如填料体积、气速等）及优化方法。读者将学习如何结合工艺需求设计高效的固定床反应器，并分析设备运行中的常见问题及解决对策。

学习目标

知识目标

（1）熟悉固定床反应器的基本结构和操作原理。

（2）掌握固定床反应器的设计方法和关键参数。

（3）了解固定床反应器的常见应用场景及限制条件。

技能目标

（1）能根据生产需求设计固定床反应器的尺寸和配置。

（2）能优化固定床反应器的操作条件以提高效率。

价值目标

（1）在固定床反应器设计中注重资源的高效利用。

（2）提升固定床反应器的安全性和环保性。

12.1 Basic Knowledge of Solid Catalysts
固体催化剂基础知识

Solid catalysts are widely used in industrial processes to accelerate chemical reactions. The catalytic process involves the interaction between the reactant molecules and the catalyst surface，resulting in a lower activation energy （更低的活化能） and faster reaction rates. Catalysts can be composed of various elements，including metals，oxides （氧化物），and zeolites （分子筛），and their composition and structure determine their catalytic function （催化作用）.

The preparation of industrial catalysts involves various techniques，including precipitation（沉淀法），impregnation（浸渍法），mechanical mixing（机械混合法），ion exchange（离子交换法），and fusion（熔融法）．These methods involve mixing the catalyst components with a carrier material（载体）to improve their dispersion and reactivity．After preparation，the catalyst is typically formed into various shapes，including pellets（压制成的颗粒），extrudates（挤出成型的颗粒），and spheres（成球型的颗粒）．

Recent advances in catalyst preparation techniques have led to the development of new methods，such as nano-technology（纳米技术），membrane catalysts（膜催化），micro-emulsion techniques（微乳液技术），physical vapor deposition（物理气相沉积）and chemical vapor deposition（化学气相沉积）．These methods offer improved catalyst performance and selectivity，as well as greater control over the catalyst's structure and properties．

In summary，solid catalysts are essential components in many industrial processes，and their preparation and shaping techniques continue to evolve to improve their performance and efficiency．The choice of catalyst and preparation method depends on the specific reaction and process requirements．The development of new catalyst preparation methods will lead to improved catalytic performance and greater control over the reaction．

技术理论

根据国际纯粹与应用化学联合会（IUPAC）于1981年提出的定义，催化剂是一种物质，它能够加速化学反应的速率而不改变该反应的标准自由焓的变化，这种作用称为催化作用。催化作用可用最简单的"假设循环"表示出来，如图12-1所示。

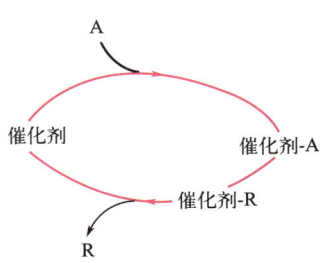

图12-1　催化反应的循环图

催化剂基本特征：①催化剂能够加快化学反应速率，但它本身并不计入化学反应的计量；②催化剂对反应具有选择性；③催化剂只能加速热力学上可能进行的化学反应，而不能加速热力学上无法进行的反应；④催化剂只能改变化学反应的速率，而不能改变化学平衡的位置（平衡常数）；⑤催化剂不改变化学平衡，意味着既能加速正反应，也能同样程度地加速逆反应，这样才能使其化学平衡常数保持不变。

绝大多数工业催化剂可分成三个组分（如图12-2所示），即活性组分、助催化剂、载体。

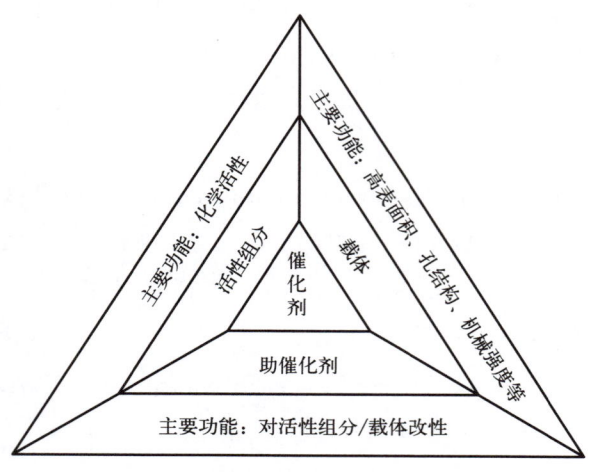

图 12-2 催化剂组分与功能关系

一种良好的催化剂不仅能选择性地催化所要求的反应，同时还必须具有一定的机械强度；有适当的形状，以使流体阻力减小并能均匀地通过；在长期使用后（包括开停车）仍能保持其活性和力学性能，即必须具备高活性、合理的流体流动性质及长寿命这三个条件。

目前，工业上使用的固体催化剂的制备方法有沉淀法、浸渍法、机械混合法、离子交换法、熔融法等。催化剂的成型方法通常有破碎成型、挤条成型、压片成型及生产球状成品的成型技术。

关键词详解

标准自由焓，standard free enthalpy，热力学函数之一，是判断等温、等压下冶金反应的方向及平衡态的依据。

化学平衡常数，chemical equilibrium constant，是指在一定温度下，可逆反应无论是从正反应或逆反应开始，不考虑反应物起始浓度大小，最后都达到平衡，这时各生成物浓度的化学计量数次幂的乘积与各反应物浓度的化学计量数次幂的乘积的比值。该值是个常数，用 K 表示。化学平衡常数一般有浓度平衡常数和压强平衡常数。

活性组分，active components，活性组分（或主催化剂）是催化剂的主要成分，是起催化作用的根本性物质。没有活性组分，就不存在催化作用。活性组分有时由一种物质组成，如乙烯氧化制环氧乙烷的银催化剂，活性组分就是银单一物质；有时则由多种物质组成，如丙烯氨氧化制丙烯腈用的钼-铋催化剂，活性组分就是由氧化钼和氧化铋两种物质组合而成。

助催化剂，promoter，一些本身对某一反应没有活性或活性很小，但添加少量于催化剂之中（一般小于催化剂总量的10％）却能使催化剂具有所期望的活性、选择性或稳定性的物质，例如，用于脱水的 Al_2O_3 催化剂以 CaO、MgO、ZnO 为助催化剂。

载体，carrier，载体是固体催化剂所特有的组分。它可以起增大表面积、提高耐热性和机械强度的作用，有时还能担当助催化剂的角色。

沉淀法，precipitation method，是借助沉淀反应，用沉淀剂（如碱类物质）将可溶性的催化剂组分（金属盐类的水溶液）转化为难溶化合物，再经分离、洗涤、干燥、焙烧、成型等工序制得成品催化剂。沉淀法是制备固体催化剂最常用的方法之一，广泛用于制备高含量的非贵金属、金属氧化物、金属盐催化剂或催化剂载体。例如采用沉淀法制备 $\gamma\text{-}Al_2O_3$ 催化剂。

浸渍法，impregnation method，是负载型催化剂最常用的制备方法。其制备步骤大体包括：①抽空载体；②载体与浸渍溶液接触；③除去过剩的溶液；④干燥；⑤煅烧及活化。例如用于加氢反应的载于氧化铝上的镍催化剂 Ni/Al_2O_3。

机械混合法，mechanical mixing method，是工业上制备多组分固体催化剂时常采用的方法。它是将几种组分用机械混合的方法制成多组分催化剂。混合的目的是促进物料间的均匀分布，提高分散度。因此，在制备时应尽可能使各组分混合均匀。尽管如此，这种单纯的机械混合，组分间的分散度不及其他方法。为了提高机械强度，在混合过程中一般要加入一定量的黏结剂。

离子交换法，ion exchange method，是利用载体表面上存在着可进行交换的离子，将活性组分通过离子交换（通常是阳离子交换）交换到载体上，然后再经过适当的后处理，如洗涤、干燥、焙烧、还原，最后得到金属负载型催化剂。常用于 Na 型分子筛及 Na 型离子交换树脂。

熔融法，melting method，是在高温条件下进行催化剂组分的熔合，使之成为均匀的混合体、合金固溶体或氧化物固溶体。在熔融温度下金属、金属氧化物都呈流体状态，有利于它们的混合均匀，促使助催化剂组分在活性组分上的分布。主要用于制备氨合成的熔铁催化剂、Fischer-Tropsch 合成催化剂、甲醇氧化的 Zn-Ga-Al 合金催化剂等。

破碎成型，crushing molding，直接将大块的固体破碎成无规则的小块。坚硬的大块物料可先用颚式破碎机，欲进一步破碎则可采用粉碎机。

挤条成型，extrusion molding，一般适用于亲水性强的物质，如氢氧化物等。将湿物料或在粉末物料中加适量的水碾捏成具有可塑性的浆状物料，然后放置 在开有小孔的圆筒中，在活塞的推动下，物料呈细条状从小孔中被挤压出来，干燥并硬化。工业上最常见的挤条成型装置是单螺杆挤条机，其结构如图12-3 所示。

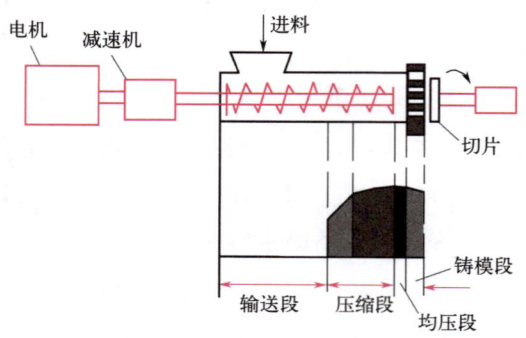

图 12-3　单螺杆挤条机结构示意图

压片成型，compression molding，压片是将粉末状物料注入圆柱形的空腔中，在空腔中的活塞上施加预定的压力，将粉压成片。

生产球状成品的成型技术，forming technology for producing spherical finished products，目前工业上常用的成型技术如下。

① 滚球法　此法适用于干燥的粉状物成型。将少量的粉末加入少量的液体（多数为水）造粒，过筛，取出一定筛分的粒子作为种子，放入滚球机中（一个斜立的可旋转的浅盘）。将待成型的粉末物料加入，并不断加入水分，由于水产生的毛细管力使粉末黏附于种子上，因而逐渐长大，成为球状物。

② 流化法　造球过程基本上与滚球法相似，但是在流化床中进行。将种子不断地加入到床层中，在床层底部将含有催化剂组分的浆料与热风一起鼓入。种子在床中处于流化状态，浆料黏附于种子上，同时逐渐干燥。由于粒子之间相互碰撞，球体颗粒逐渐长大，得到所需的球状固体催化剂。

③ 油浴法　如图12-4 所示，将可以胶凝的物料滴入（或喷入）一柱形容器中，器内盛油。由于表面张力，物料变为球状，并逐渐固化。成型后的球状产物移出容器后，即送入老化、干燥等工序。

④ 喷雾法　如图12-5 所示，对用于流化床中的微球型催化剂常可用喷雾造球法。即在一柱状容器内，将含催化剂组分的浆料自塔顶的喷头中以雾状喷入，在热风中干燥，经旋风分离器后获得产品。

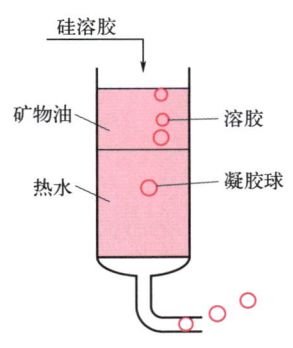

硅溶胶

矿物油 — 溶胶

热水 — 凝胶球

图 12-4　油浴造球法制 SiO_2 小球

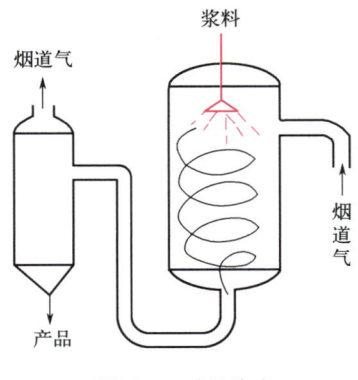

浆料

烟道气

烟道气

产品

图 12-5　喷雾造球

▶ 动画

−油浴造球
法制SiO_2
小球
−喷雾造球

互动练习

12-1　What is the purpose of solid catalysts in industrial processes?

A）To increase the activation energy of chemical reactions

B）To decrease the activation energy of chemical reactions

C）To increase the concentration of reactant molecules

D）To decrease the concentration of reactant molecules

12-2　Which of the following techniques is NOT used for preparing industrial catalysts?

A）Precipitation

B）Impregnation

C）Sedimentation

D）Mechanical mixing

12-3　What determines the catalytic function of a solid catalyst?

A）Its shape

B）Its size

C）Its composition and structure

D）Its color

12-4　What are some of the advantages of the new catalyst preparation methods?

A）Improved catalytic performance and selectivity

B）Greater control over the catalyst's structure and properties

C）Lower cost of catalyst production

D）Reduced need for carrier materials

12-5　　How are catalysts typically formed after preparation?

A）They are dissolved in a solvent

B）They are compressed into tablets

C）They are vaporized into a gas

D）They are mixed with water to form a paste

12.2 Fundamentals of Gas-Solid Catalytic Reaction Kinetics 气固相催化反应动力学基础

Gas-solid catalytic reactions play an important role in many industrial processes. The rates of these reactions are expressed by the reaction rate equations，which relate the reaction rates to the concentration of reactants and the properties of the catalyst. The reaction process involves several steps，including the adsorption of reactants（反应物吸附），the formation of intermediate species（表面反应），and the desorption of products（产物脱附）.

Physical adsorption（物理吸附）and chemical adsorption（化学吸附）are two common types of adsorption that occur during the reaction process. Physical adsorption is a weak interaction between the adsorbate（吸附剂）and the adsorbent（脱附剂），while chemical adsorption involves the formation of a chemical bond（化学键）between the two. The chemical adsorption rate（化学吸附率）can be expressed by a general rate equation，and several models have been developed to describe the adsorption process，such as the Langmuir（朗缪尔），Temkin（焦姆金）and Freundlich（弗鲁德里希），models.

The intrinsic kinetics（本征动力学）of the reaction can be described by the intrinsic kinetic equation，which is derived from the reaction rate equation. Two types of intrinsic kinetic equations are commonly used，the hyperbolic（双曲线的）intrinsic kinetic equation and the power-law（幂函数）intrinsic kinetic equation. The hyperbolic equation assumes that the reaction rate is proportional to the concentration of adsorbed species，while the power-law equation assumes that the reaction rate is proportional to the concentration of adsorbed species raised to a power.

Understanding the gas-solid catalytic reaction kinetics is crucial for optimizing the design and operation of catalytic reactors. Through the study of reaction rates and adsorption models，researchers can gain insight into the fundamental mechanisms of catalytic reactions and develop more efficient catalysts for industrial processes.

技术理论

根据化学反应速率定义式（12-1）

$$反应速率 = \frac{反应量}{反应区域 \times 反应时间} \tag{12-1}$$

式（12-1）中的反应区域，对于气固相催化反应过程有以下几种选择：

① 选用催化剂体积，反应速率 $(-r_A)'$ 单位为 $kmol/(m^3$ 催化剂 $\cdot h)$；

② 选用催化剂质量，反应速率 $(-r_A)''$ 单位为 $kmol/(kg$ 催化剂 $\cdot h)$；

③ 选用催化剂堆积体积，即反应器中催化剂床层体积，反应速率 $(-r_A)'''$ 单位为 $kmol/(m^3$ 床层 $\cdot h)$。

由此可见，即使描述同一反应过程，反应区域的选择不同，反应速率的数值大小和单位均可不同。

气固相反应本征动力学是研究不受扩散干扰条件下的固体催化剂与其相接触的气体之间的反应动力学。

一般而言，气固相催化反应过程经历以下七个步骤，如图 12-6 所示。

① 反应组分从流体主体向固体催化剂外表面传递（外扩散过程）；

② 反应组分从催化剂外表面向催化剂内表面传递（内扩散过程）；

③ 反应组分在催化剂表面的活性中心吸附（吸附过程）；

④ 在催化剂表面上进行化学反应（表面反应过程）

⑤ 反应产物在催化剂表面上脱附（脱附过程）；

⑥ 反应产物从催化剂内表面向催化剂外表面传递（内扩散过程）；

⑦ 反应产物从催化剂外表面向流体主体传递（外扩散过程）。

七个步骤中，第①和第⑦步是气相主体通过气膜与颗粒处表面进行物质传递，称为外扩散过程；第②和第⑥步是颗粒内的传质，称为内扩散过程；第③和第⑤步是在颗粒表面上进行化学吸附和化学脱附的过程；第④步是在颗粒表面上进行的表面反应动力学过程。以上七个步骤是前后串联的：

外扩散——→内扩散——→吸附——→表面反应——→脱附——→内扩散——→外扩散

表面过程

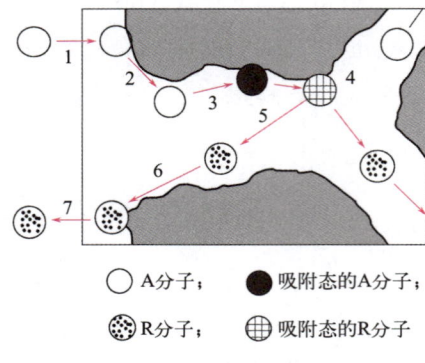

○ A分子；　● 吸附态的A分子；

⊕ R分子；　⊞ 吸附态的R分子

图 12-6　气固相催化反应过程

吸附速率方程式与吸附平衡方程式在具体应用时存在一定困难，很多学者对此提出一些简化模型，使得方程能在实践中得到应用。较著名的模型有朗缪尔吸附模型、焦姆金吸附模型和弗鲁德里希吸附模型。

关键词详解

气固相反应本征动力学，intrinsic kinetics of gas-solid reaction，在多孔催化剂上进行的气固相催化反应，由反应物在位于催化剂内表面的活性位上的化学吸附、活化吸附态组分进行化学反应和产物的脱附三个连串步骤组成。因此，气固相催化反应本征动力学的基础是化学吸附。

吸附速率方程式，adsorption rate equation，

$$r_a = k_a p_A \theta_V = A_{a0} \exp(-E_a/RT) p_A \theta_V \tag{12-2}$$

式中，r_a 为吸附速率，Pa/h；E_a 为吸附活化能，kJ/kmol；p_A 为 A 组分在气相中的分压，Pa；θ_V 为空位率；k_a 为吸附速率常数，h^{-1}；A_{a0} 为吸附指前因子，h^{-1}。

吸附平衡方程式，adsorption equilibrium equation，

$$p_A = \frac{A_{d0}}{A_{a0}} \times \frac{\theta_A}{\theta_V} \exp\left(-\frac{q}{RT}\right) \tag{12-3}$$

式中，A_{d0} 为脱附指前因子，h^{-1}，A_{a0} 为吸附指前因子，h^{-1}，θ_A 为组分 A 的覆盖率，θ_V 为空位率。

朗缪尔吸附模型，Langmuir adsorption modeling，包括以下四个基本假设：

① 催化剂表面各处的吸附能力是均匀的，各吸附位具有相同的能量；

② 被吸附物仅形成单分子层吸附；

③ 吸附的分子间不发生相互作用，也不影响分子的吸附作用；

④ 所有吸附的机理是相同的。

焦姆金吸附模型，Temkin adsorption modeling，不满足理想吸附条件的吸附，都称为真实吸附。以焦姆金和弗鲁德里希为代表提出不均匀表面吸附理论，真实吸附模型认为固体表面是不均匀的，各吸附中心的能量不等，有强有弱。吸附时吸附分子首先占据强的吸附中心，放出的吸附热大。随后逐渐减弱，放出的吸附热也愈来愈小。

弗鲁德里希吸附模型，Frieundrich adsorption modeling，认为吸附活化能、脱附活化能以及吸附热随覆盖率的不同而有差异，但弗鲁德里希吸附模型认为活化能与覆盖率之间并非线性关系，而是对数函数关系。

互动练习

12-6　What is the general expression for chemical adsorption rate in gas-solid catalytic reactions?

A）$q = k \cdot p_{gas}$

B）$q = k \cdot c_{gas}$

C）$q = k \cdot p_{gas} \cdot c_{solid}$

D）$q = k \cdot p_{gas} \cdot \theta$

12-7　Which of the following is NOT a step in the gas-solid catalytic reaction process?

A）External mass transfer

B）Chemical adsorption

C）Internal diffusion

D）Heat exchange

12-8　Which model describes adsorption by assuming a uniform distribution of adsorption energy?

A）Langmuir model

B）Freundlich model

C）Temkin model

D）BET model

12-9　What is the power function equation that describes the intrinsic kinetics of gas-solid catalytic reactions?

A）$r = k \cdot \theta^m$

B）$r = k \cdot c_{gas}^n$

C）$r = k \cdot p_{gas}^m$

D）$r = k \cdot \theta^n$

12-10　Which of the following is a type of intrinsic kinetic equation for gas-solid catalytic reactions?

A）Sinusoidal kinetic equation

B）Hyperbolic kinetic equation

C）Parabolic kinetic equation

D）Exponential kinetic equation

12.3　Design of Fixed Bed Reactors 固定床反应器设计

Fixed bed reactor design is an essential aspect of chemical engineering related to void fraction（空隙率）. The design process involves the analysis of the flow of fluid，mass transfer，and heat transfer within the reactor. The reactor's performance is influenced by the shape，size，and configuration（构造）of the catalyst particles. In designing a fixed bed reactor，the following factors must be considered.

Firstly，fluid flow through the reactor bed is determined by the catalyst particle size and shape，the average particle diameter of the mixed bed，the void fraction，and the radial velocity distribution（径向速度分布）of the bed. The pressure drop of the fluid through the bed is also an important consideration.

Secondly，mass transfer inside the reactor is essential to ensure the efficient conversion of reactants. The rate of mass transfer depends on the design of the reactor，such as the configuration and orientation（排布）of the catalyst bed.

Thirdly，heat transfer must be considered to ensure temperature control and prevent thermal runaway. Heat transfer within the reactor bed can be optimized through the selection of suitable cooling systems（冷却系统）.

The design of a fixed bed reactor can be accomplished using empirical models or mathematical models. Empirical models（经验模型）involve the use of simple equations based on the experimental data. On the other hand，mathematical models（数学模型）use differential equations to simulate the reactor's behavior.

In conclusion，fixed bed reactor design is a complex process that requires a detailed understanding of fluid flow，mass transfer，and heat transfer. An efficient reactor design can be achieved by optimizing the catalyst particle size and shape，bed void fraction，and radial velocity distribution.

技术理论

固定床反应器中的流体流动可以用简单的扩散模型进行模拟，即认为流动由两部分组成：一部分为流体以平均流速沿轴向作理想置换式的流动；另一部分为流体的径向和轴向的混合扩散，包括分子扩散（层流时为主）和涡流扩散（湍流时为主）。根据不同的混合扩散程度，将两个部分叠加。

流体流过固定床层的压降，主要是由流体与颗粒表面间的摩擦阻力和流体在孔道中的收缩、扩大和再分布等局部阻力引起。当流动状态为滞流时，以摩擦阻力为主；当流动状态为湍流时，以局部阻力为主。

固定床反应器中的传质过程包括外扩散、内扩散和床层内的混合扩散。固定床的传热实质上包括了颗粒内的传热、颗粒与流体之间的传热以及床层与器壁的传热等几个方面。

固定床反应器的设计计算，一般包括催化剂用量、反应器床层高度和直径、传热面积及床层压降的计算等。主要有经验法和数学模型法。

气固相催化反应固定床反应器的计算通常包括下述三种情况：

① 为了完成一定生产任务，对反应器进行工艺设计。即已知原料气进反应器时的各项参数（温度、压力、流量及组成），并确定了反应器出口气体的组成（或转化率）时，通常计算求出反应器的直径，催化剂床层高度以及有关工艺参数。

② 对已有的反应器（已知其直径和催化剂层高度），在规定了原料气各项参数后，计算该反应器能否实现工艺指标（即反应器出口气体是否符合要求）。

③ 对已有的反应器，为满足生产所需要的产品要求，计算反应器的最大生产能力（即反应器可以处理的最大原料量）。

上述三种任务中，第一种为设计任务，后两种为反应器校核。尽管要求不同，但计算原理和方法是相同的。

关键词详解

层流，laminar flow，是流体的一种流动状态，它作层状的流动。流体在管内低速流动时呈现为层流，其质点沿着与管轴平行的方向作平滑直线运动。流体的流速在管中心处最大，其近壁处最小。

涡流扩散，eddy diffusion，是指流体在床层内形成旋涡运动时，颗粒与流体之间的扩散现象，对于床层内流体的混合和传质传热有重要影响。

湍流，turbulence，是流体的一种流动状态。当流速很小时，流体分层流动，互不混合，称为层流，或称为片流；逐渐增加流速，流体的流线开始出现波状的摆动，摆动的频率及振幅随流速的增加而增加，此种流况称为过渡流；

当流速增加到很大时，流线不再清楚可辨，流场中有许多小漩涡，称为湍流，又称为乱流、扰流或紊流。

相当直径，equivalent diameter，是指固定床反应器中颗粒或催化剂的等效直径，用于描述床层内颗粒的尺寸特征。主要包括体积相当直径、面积相当直径和比表面积相当直径。

① 体积相当直径 d_V　即采用体积相同的球形颗粒直径来表示非球形颗粒直径。

$$d_V = (6V_P/\pi)^{\frac{1}{3}} \tag{12-4}$$

式中，V_P 为非球形颗粒的体积，m^3。

② 面积相当直径 d_a　即采用外表面积相同的球形颗粒直径来表示非球形颗粒直径。

$$d_a = (A_P/\pi)^{\frac{1}{2}} \tag{12-5}$$

式中，A_P 为非球形颗粒的外表面积，m^2。

③ 比表面积相当直径 d_S　即采用比表面积相同的球形颗粒直径来表示非球形颗粒的直径。

非球形颗粒的比表面积为 $S_V = A_P/V_P$，比表面积等于 S_V 的球形颗粒有如下关系式

$$S_V = \pi d_S^2 \left/ \left(\frac{1}{6}\pi d_S^3 \right) \right. = 6/d_S \tag{12-6}$$

所以　　　　　　　　　$$d_S = 6/S_V = 6V_P/A_P \tag{12-7}$$

在固定床的流体力学研究中，非球形颗粒的直径常常采用体积相当直径。在传热传质的研究中，常常采用面积相当直径。

④ 形状系数 ϕ_S　非球形颗粒的外表面积一定大于等体积的圆球的外表面积。因此，引入一个无量纲系数，称为颗粒的形状系数 ϕ_S，其值如下

$$\phi_S = A_S/A_P \tag{12-8}$$

式中，A_S 为与非球形颗粒等体积圆球的外表面积，m^2；$A_S = \pi d_V^2$。

ϕ_S 即与非球形颗粒体积相等的圆球的外表面积与非球形颗粒的外表面积之比。对于球形颗粒，$\phi_S = 1$；对于非球形颗粒，$\phi_S < 1$。形状系数说明了颗粒与圆球的差异程度。

三种相当直径用 ϕ_S 联系起来，有如下关系

$$d_S = \phi_S d_V = \phi_S^{\frac{3}{2}} d_a \tag{12-9}$$

当催化剂床层由大小不一、形状各异的颗粒组成时，就有一个如何计算混合颗粒的平均粒度及形状系数的问题。

对于大小不等的混合颗粒，如果颗粒不太细（大于 0.075mm），平均直径可由筛分分析数据来决定。将混合颗粒用标准筛组进行筛分，分别称量留在各号筛上的颗粒质量，然后根据颗粒的总质量分别算出各种颗粒所占的分数。

在某一号筛上的颗粒，其直径 d_i 通常为该号筛孔净宽及上一号筛孔净宽的几何平均值（即两相邻筛孔净宽乘积的平方根）。如混合颗粒中，直径为 d_1、d_2、$\cdots$、d_n 的颗粒的质量分数分别为 x_1、x_2、$\cdots$、x_n，则混合颗粒的平均直径用算术平均直径法计算为

$$\overline{d_P} = \sum_{i=1}^{n} x_i d_i \tag{12-10}$$

若以调和平均法计算，则

$$\frac{1}{d_P} = \sum_{i=1}^{n} \frac{x_i}{d_i} \tag{12-11}$$

在固定床和流化床的流体力学计算中，用调和平均直径较为符合实验数据。

空隙率，void fraction，是指固定床反应器中床层空隙体积与总体积之比，反映了颗粒与颗粒之间的间隙。

$$\varepsilon = 1 - \frac{\rho_b}{\rho_s} \tag{12-12}$$

式中，ε 为床层空隙率；ρ_b 为催化剂床层堆积密度，即单位体积催化剂床层具有的质量，kg/m^3；ρ_s 为催化剂的表观密度，即单位体积催化剂颗粒具有的质量，kg/m^3。

床层空隙率 ε 的大小与下列因素有关：颗粒形状、颗粒的粒度分布、颗粒表面的粗糙度、充填方式、颗粒直径与容器直径之比等。

床层压降，pressure drop，流体通过固体床层时由于阻力引起的压力损失，是影响固定床反应器操作和性能的重要因素之一。

计算压降的公式很多，常用的一个是仿照流体在空管中流动的压降公式而导出的埃冈（Ergun）公式。

固定床的压降可表示为

$$\Delta p = f_M \frac{\rho_f u_0^2}{d_S} \times \frac{1-\varepsilon}{\varepsilon^3} L \tag{12-13}$$

式中，Δp 为压降，Pa；f_M 为修正摩擦系数；L 为管长，m；ρ_f 为流体密度，kg/m^3；u_0 为流体空床平均流速，即以床层空截面积计算的流体平均流速，m/s；d_S 为催化剂颗粒的比表面积相当直径；ε 为床层空隙率。

经实验测定，修正摩擦系数 f_M 与修正雷诺数 Re_M 的关系可表示如下：

$$f_M = \frac{150}{Re_M} \times \frac{1-\varepsilon}{\varepsilon^3} L \qquad (12-14)$$

$$Re_M = \frac{d_S}{Re_M} + 1.75 \qquad (12-15)$$

$$Re_M = \frac{d_S \rho_f u_0}{\mu_f} \times \frac{1}{1-\varepsilon} = \frac{d_S G}{\mu_f} \times \frac{1}{1-\varepsilon} \qquad (12-16)$$

式中，μ_f 为流体的黏度，Pa·s；G 为流体的质量流速，kg/(m²·s)。

当 $Re_M < 10$ 时，流体处于层流状态，式（12-14）中 $\frac{150}{Re_M} \geqslant 1.75$，即式（12-13）可简化为

$$\Delta p = 150 \frac{\mu_f u_0}{d_S^2} \times \frac{(1-\varepsilon)^2}{\varepsilon^3} L \qquad (12-17)$$

当 $Re_M > 10$ 时，流体处于湍流状态，式（12-14）中 $\frac{150}{Re_M} \leqslant 1.75$，即式（12-13）可简化为

$$\Delta p = 1.75 \frac{\rho_f u_0^2}{d_S^2} \times \frac{1-\varepsilon}{\varepsilon^3} L \qquad (12-18)$$

如果床层中催化剂颗粒大小不一，用式（12-17）、式（12-18）时，应采用颗粒的平均相当直径 $\overline{d_S}$。

$\overline{d_S}$ 可按下式计算

$$\overline{d_S} = \frac{6}{\sum x_i S_{Vi}} = \frac{1}{\sum \left(\frac{x_i}{d_{Si}} \right)} \qquad (12-19)$$

式中，d_S 为平均比表面积相当直径，m；x_i 为颗粒 i 筛分所占的体积分数（如果各筛分颗粒的密度相同，则体积分数亦为质量分数）；S_{Vi} 为颗粒 i 筛分的比表面积，m²/m³。

如果各种粒度颗粒的形状系数相差不大，$\overline{d_S}$ 即为按式（12-11）计算的调和平均直径 $\overline{d_P}$ 与平均形状系数的乘积。

影响床层压降的因素可分为两类：一类来自流体，如流体的黏度、密度等物理性质和流体的质量流速；另一类来自床层，如床层的高度、空隙率和颗粒的物理特性（如粒度、形状和表面粗糙度等）。

经验法，empirical method，是取用实验室、中间试验装置或工厂现有生产装置中最佳条件下测得的一些数据，如空速、催化剂空时收率及催化剂负荷等作为工艺计算的依据。

经验法工艺计算的前提是新设计计算的反应器也能保持与提供数据的装置

相同的操作条件，如催化剂性质、粒度、原料组成、流体流速、温度和压力等。由于规模的改变，要做到全部相同是困难的，尤其是温度条件。因此这种方法虽能在缺乏动力学数据的情况下简单方便地估算出催化剂体积，但因对整个反应系统的反应动力学、传质、传热等特性缺乏真正的了解，是比较原始的、不精确的，不能实现高倍数的放大。

数学模型法，numerical modeling method，20 世纪中期发展起来的先进方法，它建立在对反应器内全部过程的本质和规律有一定认识的基础上，用数学方程式来比较真实地描述实际过程—即建立过程的数学模型，运用计算机可以进行高倍数放大的工艺计算。当然，数学模型的可靠性和基础物性数据测定的准确性是正确计算的关键。在讨论固定床反应器内流体流动、传热和传质过程的基础上，可以建立固定床反应器内传递过程的数学模型，结合反应动力学的数学模型，就能得到描述固定床反应器内全部过程的数学模型。目前，固定床反应器的数学模型被认为是反应器中比较成熟可靠的模型。它不仅用于设计计算，也用于检验现有反应器的操作性能，以探求技术改造的途径和实现最优控制。

催化反应器的数学模型，根据反应动力学可分为非均相与拟均相两类，根据催化床中温度分布可分为一维模型和二维模型，根据流体的流动状况又可分为理想流动模型（包括理想置换和理想混合流动模型）和非理想流动模型。

绝大部分工业催化反应的传质和传热过程对宏观速率都有影响。例如烃类蒸气转化的催化剂要同时考虑气流主体与催化剂颗粒外表面的相间传质和传热及颗粒内部的传质和传热。把这些传递过程对反应速率的影响计入模型，称为"非均相"模型。

如果反应属于化学动力学控制，催化剂颗粒外表面上及颗粒内部反应组分的浓度及温度都与气流主体一致，计算过程与均相反应一样，故称为"拟均相"模型。如果某些催化过程的宏观动力学研究得不够，只能按本征动力学处理，而将传递过程的影响、催化剂的中毒、结焦、衰老、还原等项因素合并成为"活性校正系数"，这种处理方法属于"拟均相"模型。应注意活性校正系数与本征动力学参数、催化剂粒度、反应器结构、催化剂装载于反应器中的位置、毒物的品种及含量、催化剂的还原情况及使用时间等条件有关。

若只考虑反应器中沿着气流方向的浓度差和温度差，称为"一维模型"；若同时计入垂直于气流方向的浓度差和温度差，称为"二维模型"。一维拟均相理想置换流动模型是最基础的模型，在此基础上，按各种类型反应器的实际情况，计入轴向返混、径向浓度差及温度差、相间及颗粒内部的传质和传热，便形成了表 12-1 的分类。

表 12-1　催化反应器数学模型分类

分类		A. 拟均相		B. 非均相
一维	AⅠ	理想流动基础模型	BⅠ	AⅠ＋相间及粒内浓度分布及温度分布
	AⅡ	AⅠ＋轴向返混	BⅡ	BⅠ＋轴向返混
二维	AⅢ	AⅡ＋径向混合	BⅢ	BⅡ＋径向混合

表 12-1 中基础模型的数学表达式最简单，所需的模型参数最少，数学运算也最简单。模型中考虑的问题越多，所需的传递过程参数越多，其数学表达式越复杂，求解时也越费时。处理具体问题时，一定要针对具体反应过程及反应器的特点进行分析，选用合适的模型。如果通过检验认为可以进行合理的假设而选用简化模型时，则采用简化模型进行模拟计算和模拟放大。具体可参考相关资料手册。

互动练习

12-11　Which factors affect the fluid flow inside a fixed bed reactor?

A）Catalyst particle diameter and shape factor，average diameter and shape factor of mixed particles，bed void fraction and radial flow velocity distribution

B）Chemical reactions and heat transfer rate

C）Reactor size and shape

D）Catalyst activity and selectivity

12-12　What is the purpose of designing a fixed bed reactor?

A）To increase the pressure drop in the reactor

B）To optimize the reaction rate and yield

C）To reduce the cost of the catalyst

D）To simplify the reactor operation

12-13　Which method can be used to design a fixed bed reactor?

A）Empirical method only

B）Mathematical modeling method only

C）Both empirical and mathematical modeling methods

D）None of the above

12-14　What is the role of bed void fraction in a fixed bed reactor?

A）To increase the pressure drop in the reactor

B）To reduce the flow rate of the reactants

C）To optimize the heat transfer rate

D）To provide space for the catalyst particles and allow the flow of reactants and products

12-15　What is the effect of radial flow velocity distribution on the performance of a fixed bed reactor?

A）It has no effect on the performance of the reactor

B）It can lead to uneven reaction rates and incomplete conversion

C）It increases the yield of the desired product

D）It decreases the yield of the desired product

任务13

Design of Fluidized Bed Reactors
流化床反应器设计

任务要点

本任务详细介绍流化床反应器的设计原则和优化技术，涵盖气速、压降和传质效率等核心参数。读者将学习如何根据工艺需求设计高效的流化床反应器，并掌握提升传质效率和减少能耗的方法。

学习目标

知识目标

（1）了解流化床反应器的基本原理及特性。

（2）掌握流化床反应器的设计要点，包括气速和压降计算。

（3）了解流化床反应器在工业中的典型应用。

技能目标

（1）能设计和评估流化床反应器的性能。

（2）能解决流化床反应器运行中常见的问题。

价值目标

（1）提高流化床反应器设备运行的稳定性和可靠性。

（2）注重流化床反应器设计中的经济性和环保性。

13.1 Basic Concepts of Fluidization
流态化基本概念

Fluidization（流态化）is a phenomenon where a solid particle bed behaves like a liquid due to the fluid flow of gas through it. There are two types of fluidized bed reactors：the bubbling fluidized bed（BFB）（鼓泡流化床）and the circulating fluidized bed（CFB）（循环流化床）.

BFB has a lower gas velocity（气速）than CFB，and the gas flow causes the bed to bubble like boiling water. CFB，on the other hand，has a higher gas velocity that causes the particles to circulate and creates a more uniform distribution of temperature and reactants. The pressure drop and gas velocity in a fluidized bed are related to the particle size，shape，and density，as well as the bed height and gas properties. The bubbles in a fluidized bed rise to the surface and burst（破裂），causing particle mixing and heat transfer. In a BFB，the bubbles are larger and fewer，while in a CFB，the bubbles are smaller and more numerous. Understanding the behavior of gas and particles in fluidized beds is critical in the design and operation of fluidized bed reactors in various industrial processes.

技术理论

固体流态化是指将固体颗粒通过气体或液体介质使其呈现流动状态的物理过程。在固体流态化过程中，颗粒之间的相互作用力被气体（或液体）介质所克服，从而形成流态。当气体通过固体颗粒床层时，随着气速的改变，分别经历固定床、临界流化床、流化床和气流输送四个阶段。

不同的流体，固体流态化现象有所不同，可分为散式流态化和聚式流态化。

求出临界流化速度和颗粒带出速度，原则上确定流化床操作速度的范围，其范围较宽，要最终确定操作速度，还必须考虑许多因素，加以综合分析比较，才能得出适当的选择。

作为反应器的流化床，其中的流体流动及传递过程是非常复杂的，并且气体和颗粒在床内的混合是不均匀的。如图 13-1 所示，气体经分布板进入床

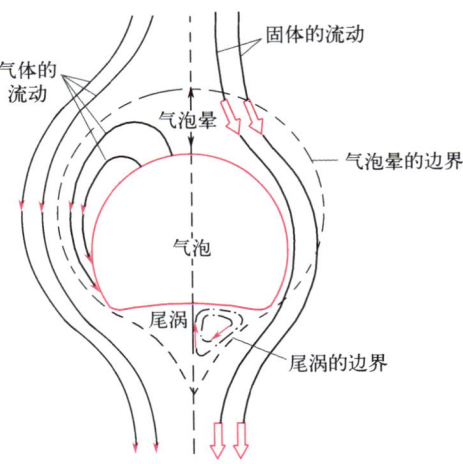

图 13-1　气泡及其周围气体与颗粒运动情况

层后，一部分与固体颗粒混合构成乳化相，另一部分不与固体颗粒混合而以气泡状态在床层中上升，这部分气体构成气泡相。气泡在上升中，因聚并和膨胀而增大，同时不断与乳化相间进行质量交换，即将反应物组分传递到乳化相中，使其在催化剂上进行反应，又将反应生成的产物传到气泡相中来。

关键词详解

固定床、临界流化床、流化床和气流输送四个阶段，four stages：fixed bed，critical fluidized bed，fluidized bed，and airflow conveying，当流速较低时，固体颗粒静止不动，颗粒之间仍保持接触，床层的空隙率及高度都不变，流体只在颗粒间的缝隙中通过，此时属于固定床，如图 13-2（a）所示。继续增大流速，当流体通过固体颗粒产生的摩擦力与固体颗粒的浮力之和等于颗粒自身重力时，颗粒位置开始有所变化，床层略有膨胀，但颗粒还不能自由运动，颗粒间仍处于接触状态，此时称为初始或临界流化床，如图 13-2（b）所示。当流速进一步增加到高于初始流化的流速时，颗粒全部悬浮在向上流动的流体中，即进入流化状态。如果床层下部进入的流体是气体，流化床阶段气体以鼓泡的方式通过床层。随着气体流速的继续增加，固体颗粒在床层中的运动也越激烈，此时的气固系统具有类似于液体的特性。随着容器形状变化，床层高度发生变化，但有明显的上界面，这时的床层称为流化床，如图 13-2（c）所示。当气流速度升高到某一极限值时，流化床上界面消失，颗粒分散悬浮在气流中，被气流带走，这种状态称为气流输送或稀相输送床，如图 13-2（d）所示。图 13-3 展示了流化床流速变化引起的压降的变化。

动画

不同流速时床层的变化

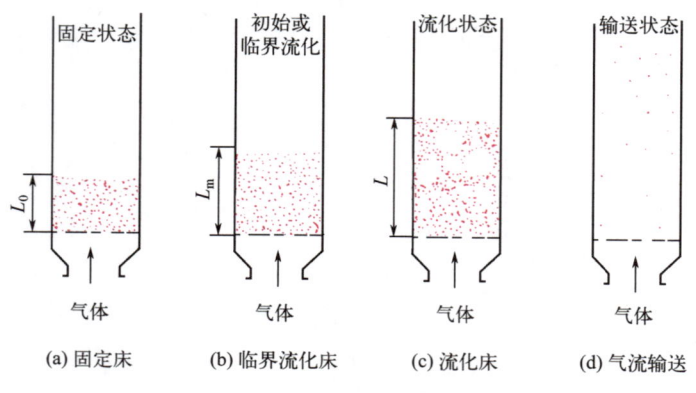

(a) 固定床　　(b) 临界流化床　　(c) 流化床　　(d) 气流输送

图 13-2 不同流速时床层的变化

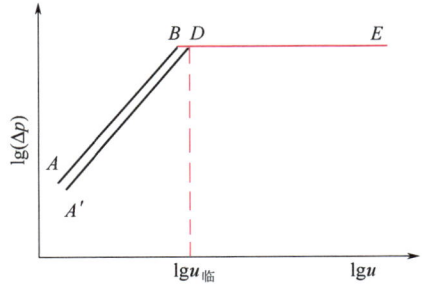

图 13-3　流化床压降-流速关系

　　散式流态化，particulate fluidization，不同的流体，固体流化现象也不同。据此一般可分为聚式流化床和散式流化床。对于液固系统，当流速高于最小流化速度时，随着流速的增加，得到的是平稳的、逐渐膨胀的床层，固体颗粒均匀地分布于床层各处，床面清晰可辨，略有波动，但相当稳定，床层压降的波动也很小或基本保持不变。即使在流速较大时，也看不到鼓泡或不均匀的现象。这种床层称为散式流化床，或均匀流化床、液体流化床，如图 13-4（a）所示。

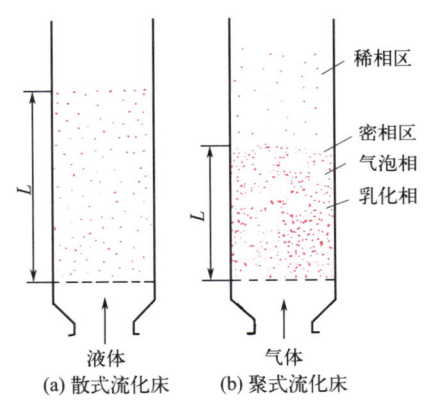

　　聚式流态化，aggregative fluidization，当流体为气体时，即气固系统的流化床中，气体流速超过临界流化速度以后，有相当一部分气体以气泡形式通过床层，气泡在床层中上升并相互聚并，引起床层的波动，这种波动随流速的增大而增大。同时床面也有相应的波动，波动剧烈时很难确定其具体位置，这与液固系统中的清晰床面大不相同。由于床内存在气泡，气泡向上运动时将部分颗粒夹带至床面，到达床面时气泡发生破裂，这部分颗粒由于自身重力作用又落回床内。整个过程中气泡不断产生和破裂，所以气固流化床的外观与液固流化床不同，颗粒不是均匀地分散于床层中，而是程度不同地一团一团聚集在一起作不规则的运动。在固体颗粒粒度比较小时，这种现象更为明显。因此，气固系统的这种流化床称为聚式流化床，如图 13-4（b）所示。

图 13-4　流化床的类型

　　临界流化速度，critical fluidize velocity，也称起始流化速度、最低流化速度，是指颗粒层由固定床转为流化床时流体的表观速度，用 u_{mf} 表示。

对于小颗粒 $$u_{mf}=\frac{d_p^2(\rho_p-\rho_f)g}{1650\mu_f}\quad(Re<20)\tag{13-1}$$

对于大颗粒 $$u_{mf}^2=\frac{d_p(\rho_p-\rho_f)g}{24.5\rho_f}\quad(Re>1000)\tag{13-2}$$

采用上述各式计算时，应将所得 u_{mf} 值代入 $Re = d_p u_{mf} \rho_f / \mu_f$ 中，检验其是否符合规定的范围。如不相符，应重新选择公式计算。

$$u_{mf} = 0.00923 \frac{d_p^{1.82}(\rho_p - \rho_f)^{0.94} g}{\mu_f^{0.88} \rho_f^{0.06}} \quad (m/s) \tag{13-3}$$

式（13-3）只适用于 $Re < 10$，即较细的颗粒。如果 $Re > 10$，则需要再乘以图 13-4 中的校正系数。图 13-5 展示了 $Re > 10$ 时的校正系数。

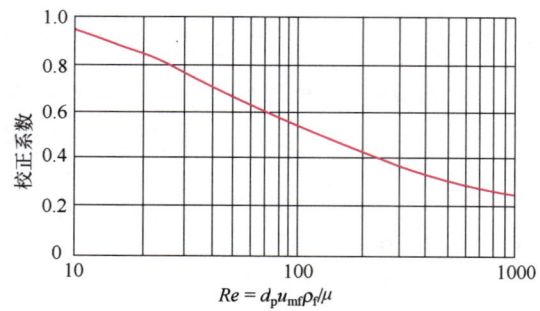

图 13-5　Re> 10 时的校正系数

颗粒带出速度，particle extraction speed，是流化床中流体速度的上限，也就是气速增大到此值时流体对粒子的曳力与粒子的重力相等，粒子将被气流带走。这一带出速度，或称终端速度，近似地等于粒子的自由沉降速度。用 u_t 表示。

$$u_t = \frac{d_p^2 (\rho_p - \rho_f) g}{18 \mu_f} \quad (Re_t < 0.4) \tag{13-4}$$

$$u_t = \left[\frac{4}{225} \frac{(\rho_p - \rho_f)^2 g^2}{\rho_f \mu_f} \right]^{\frac{1}{3}} d_p \quad (0.4 < Re_t < 500) \tag{13-5}$$

$$u_t = \left[\frac{3.1 d_p (\rho_p - \rho_f) g}{\rho_f} \right]^{\frac{1}{2}} \quad (500 < Re_t < 2 \times 10^5) \tag{13-6}$$

采用上列诸式计算的 u_t 也需再代入 Re_t 中以检验其范围是否相符。

气泡相，bubble phase，流化床反应器中形成的由气体和固体颗粒组成的气-固两相流动状态。气泡相的形成是由于气体通过床层时，使颗粒悬浮并形成气泡。

乳化相，emulsion phase，流化床反应器中形成的由气体和液体组成的气-液两相流动状态。在乳化相中，气体通过搅拌或湍流引起液滴的分散，从而增加了气体和液体之间的质量传递。

13-1　What is the main topic of the passage?

A）Fluidization basics

B）Solid-state catalysis

C）Reactor design

D）Kinetics of gas-solid reactions

13-2　What are the two types of fluidized beds mentioned in the passage?

A）Gas-phase and liquid-phase beds

B）Fixed and moving beds

C）Dispersed and aggregated beds

D）Shallow and deep beds

13-3　What is the main difference between dispersed and aggregated fluidized beds?

A）Dispersed beds have a lower pressure drop than aggregated beds

B）Aggregated beds have a more uniform fluid distribution than dispersed beds

C）Dispersed beds have a more stable fluidization behavior than aggregated beds

D）Aggregated beds have larger bubbles and a more chaotic fluidization behavior than dispersed beds

13-4　What are the two types of fluidizations mentioned in the passage?

A）Gas and liquid fluidization

B）Pressure and flow rate fluidization

C）Dense and dilute phase fluidization

D）Radial and axial fluidization

13-5　What is the significance of the pressure drop in a fluidized bed?

A）It determines the size of the bubbles in the bed

B）It affects the distribution of the fluid within the bed

C）It determines the amount of catalyst required for the reaction

D）It affects the heat transfer efficiency within the bed

13.2 Mass Transfer in Fluidized Bed Reactors 流化床反应器传质

Fluidized bed reactors are widely used in chemical processing because of their excellent mass and heat transfer capabilities.

The mass transfer in fluidized beds involves the exchange of molecules between the fluid and solid phases. In this process，the rate of transfer is affected by the concentration gradients（浓度梯度）between the two phases，the properties of the fluid and solid，and the flow conditions. The presence of gas bubbles and emulsion phase（乳化相）in fluidized beds can enhance the mass transfer rate. Therefore，proper design and operation of fluidized bed reactors can optimize the mass transfer rate and improve the efficiency of the process.

13.3 Heat Transfer in Fluidized Bed Reactors 流化床反应器传热

The heat transfer in fluidized bed reactors involves three modes：heat transfer between particles，heat transfer between gas and solid particles，and heat transfer between the bed and the reactor walls（反应器壁）or heat exchanger surfaces（换热器表面）. Efficient heat transfer is crucial for maximizing the reaction rate and improving the overall efficiency of the reactor. The heat transfer rate in fluidized bed reactors can be enhanced by adjusting the fluidization parameters（流化参数），such as bed temperature，gas velocity，and particle size distribution，as well as optimizing the reactor design.

技术理论

流化床反应器传质包括颗粒与流体间的传质、气泡与乳化相间的传质。

流化床反应器传热包括颗粒与颗粒之间的传热、相间即气体与固体颗粒之间的传热、床层与内壁间的和床层与浸没于床层中的换热器表面间的传热。在一般情况下，自由流化床是等温的，粒子与流体之间的温差，除特殊情况外，可以忽略不计。重要的是床层与内壁间和床层与浸没于床层中的换热器表面间的传热。

关键词详解

自由流化床，freedom fluidized bed，流化床筒体内不设内部构件，为自由床。

互动练习

13-6　What is the primary mechanism of mass transfer in fluidized bed reactors?

A）Convection

B）Diffusion

C）Radiation

D）Conduction

13-7　How do gas bubbles and emulsified phases affect mass transfer in fluidized bed reactors?

A）They reduce the mass transfer rate

B）They have no effect on the mass transfer rate

C）They enhance the mass transfer rate

D）They change the mechanism of mass transfer

13-8　What are the three modes of heat transfer in fluidized bed reactors?

A）Radiation，convection，and conduction

B）Convection，conduction，and diffusion

C）Conduction，diffusion，and radiation

D）Heat transfer between particles，heat transfer between gas and solid particles，and heat transfer between the bed and the reactor walls or heat exchange surfaces

13-9　How can the heat transfer rate in fluidized bed reactors be enhanced?

A）By increasing the bed temperature

B）By decreasing the gas velocity

C）By increasing the particle size distribution

D）By adjusting the fluidization parameters，such as bed temperature，gas velocity，and particle size distribution，as well as optimizing the reactor design

13-10　Why is efficient heat transfer crucial for fluidized bed reactors?

A）It improves the overall efficiency of the reactor

B）It reduces the reaction rate

C）It has no effect on the reactor efficiency

D）It enhances the mass transfer rate

13.4 Design of Fluidized Bed Reactors
流化床反应器设计

The design of fluidized bed reactors involves several important factors.

The diameter and height of the reactor are two essential parameters that need to be determined based on the process requirements，as they affect the fluidization characteristics and overall performance.

The pressure drop across the bed is another critical parameter that needs to be calculated accurately to optimize the reactor performance. This involves considering the gas distribution plate，which is an important component of the reactor，and its construction and operation parameters. The pressure drop across the gas distribution plate is a critical factor that affects the fluidization of the bed.

The mathematical models of fluidized bed reactors are also important tools in designing and optimizing the reactor performance. The two-phase model（两相模型）and bubbling bed model（鼓泡床模型）are two commonly used mathematical models that can be used to simulate the behavior of the reactor and optimize its performance.

技术理论

流化床反应器设计首先是选型，再就是确定床径和床高、内部构件，并计算压降等。流化床反应器的压降主要包括气体分布板压降、流化床压降和分离设备压降。具体选型主要应根据工艺过程的特点来考虑，即化学反应的特点、颗粒或催化剂的性能、对产品的要求以及生产规模。

流化床反应器的数学模型包括两相模型、三相模型和四区模型，其中研究较多的是两相模型和鼓泡床模型。

关键词详解

床径和床高，bed diameter and bed height

（1）流化床直径

当生产规模确定后，通过物料衡算得出通过床层的总气量 Q（标准状态）。用前面介绍的方法，根据反应要求的温度、压力和气固物性确定操作气速 u，则

$$Q = \frac{1}{4}\pi D_R^2 u \times 3600 \times \frac{273}{T} \times \frac{p}{1.013 \times 10^5} \tag{13-7}$$

$$D_R = \sqrt{\frac{4 \times 1.013 \times 10^5 TQ}{273 \times 3600 \pi u p}} = \sqrt{\frac{4.132TQ}{982800\pi u p}} \tag{13-8}$$

式中，Q 为气体（标准状态）的体积流量，m^3/h；D_R 为反应器直径，m；T、p 为反应时的绝对温度（K）和绝对压力（Pa）；u 为以 T、p 计的表观气速，m/s（一般取 1/2 床高处的 p 进行计算）。

为了尽量减少气体中带出的颗粒，一般流化床反应器上部设置扩大段，扩大段直径由不允许吹出粒子的最小颗粒直径来确定。首先根据物料的物性参数与操作条件计算出此颗粒的自由沉降速度，然后按下式计算出扩大段直径 D_L。

$$Q = \frac{1}{4}\pi D_R^2 u_t \times 3600 \times \frac{273}{T} \times \frac{p}{1.013} \tag{13-9}$$

$$D_L = \sqrt{\frac{4 \times 1.013 \times TQ}{273 \times 3600 \pi u_t p}} \tag{13-10}$$

（2）流化床床高

一台完整的流化床反应器高度包括流化床高度、扩大段高度和分离高度。而流化床高度又包括临界流化床高度 L_{mf}、流化床高度 L_f 与稳定段高度 L_D。

$$L_{mf} = \frac{4W_F\tau}{\pi D_R^2 \rho_p(1 - \varepsilon_{mf})} \tag{13-11}$$

已知 L_{mf} 后，可根据床层膨胀比 R 求出流化床的床高度 L_f。床层的膨胀比定义为：

$$R = L_f/L_{mf} = (1 - \varepsilon_{mf})/(1 - \varepsilon_m) = \rho_{mf}/\rho_m \tag{13-12}$$

式中，ρ_{mf} 和 ρ_m 分别为临界流化状态和实际操作条件下床层的平均密度。

$$L_f = RL_{mf} \tag{13-13}$$

由于气固系统的不稳定性，床面有一定的起伏，为使床层稳定操作，一般在反应器计算时要考虑在床高之上增加一段高度，使之能够适应床面的起伏，这一段高度称为稳定段高度，用 L_D 表示。它主要取决于床层的稳定性和操作中浓相床层的高度变化范围。

气体分布板，gas distributor plate，工业生产用的气体分布板的型式很多，主要有密孔板、直流式、侧流式和填充式分布板、旋流式喷嘴和分枝式分布器等，每一种类型又有多种不同的结构。

密孔板又称烧结板，被认为是气体分布均匀、初生气泡细小、流态化质量最好的一种分布板。但因其易被堵塞，并且堵塞后不易排出，加上造价较高，所以在工业中较少使用。

直流式分布板结构简单，易于设计制造，但气流方向正对床层，易使床层形成沟流，小孔易于堵塞，停车时又易漏料。所以，除特殊情况外，一般不使用直流式分布板。图 13-6 所示的是直流式分布板的三种结构。

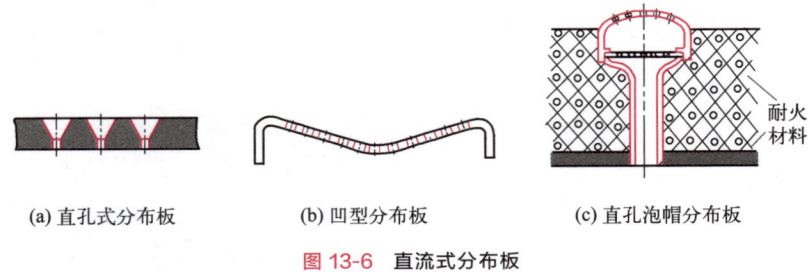

(a) 直孔式分布板　　　　(b) 凹型分布板　　　　(c) 直孔泡帽分布板

图 13-6　直流式分布板

填充式分布板是在多孔板（或栅板）和金属丝网上间隔地铺上卵石、石英砂、卵石，再用金属丝网压紧，如图 13-7 所示。其结构简单，制造容易，并能达到均匀布气的要求，流态化质量较好。但在操作过程中，固体颗粒一旦进入填充层就很难被吹出，容易造成烧结。另外，经过长期使用后，填充层常有松动，造成移位，降低了布气的均匀程度。

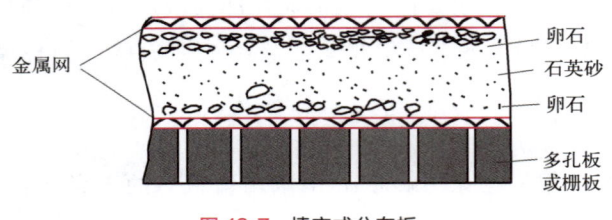

图 13-7　填充式分布板

侧流式分布板如图 13-8 所示，它是在分布板孔中装有锥形风帽，气流从锥帽底部的侧缝或锥帽四周的侧孔流出，是应用最广、效果较好的一种分布板。其中侧缝式锥帽因其不会在顶部形成小的死区，气体紧贴分布板板面吹出，适当气速下也可以消除板面上的死区，从而大大改善床层的流态化质量，避免发生烧结和分布板磨蚀现象，因此应用更广。

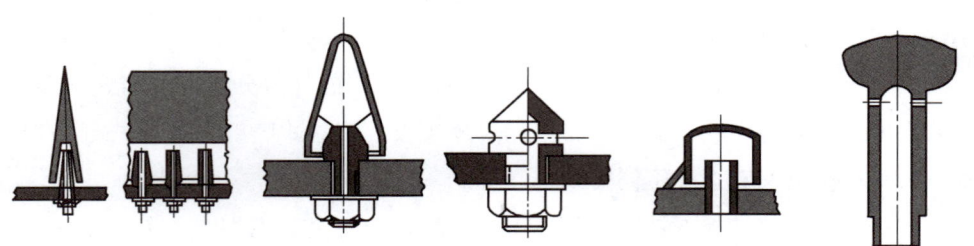

(a) 条形侧缝分布板　(b) 锥形侧缝分布板　(c) 锥形侧孔分布板　(d) 泡帽侧缝分布板　(e) 泡帽侧孔分布板

图 13-8　侧流式分布板

无分布板的旋流式喷嘴如图 13-9 所示。气体通过六个方向上倾斜 10°的喷嘴喷出，托起颗粒，使颗粒激烈搅动。中部的二次空气喷嘴均偏离径向 20°～25°，造成向上旋转的气流。这种流态化方式一般应用于对气体产品要求不严的粗粒流态化床中。

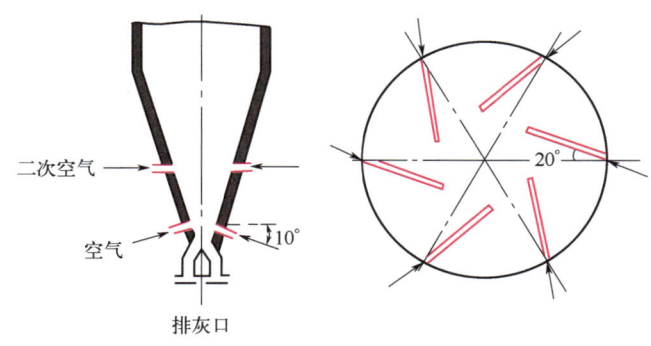

图 13-9　无分布板的旋流式喷嘴

短管式分布板是在整个分布板上均匀设置若干根短管，每根短管下部有一个气体流入的小孔，如图 13-10 所示。孔径为 9～10mm，为管径的 1/4～1/3，开孔率约 0.2%。短管长度约为 200mm。短管及其下部的小孔可以防止气体涡流，有利于均匀布气，使流化床操作稳定。

多管式气流分布器是近年来发展起来的一种新型分布器，由一个主管和若干带喷射管的支管组成，如图 13-11 所示。由于气体向下射出，可消除床层死区，也不存在固体泄漏问题，并且可以根据工艺要求设计成均匀布气或非均匀布气的结构。另外分布器本身无需同时支撑床层质量，可做成薄型结构。

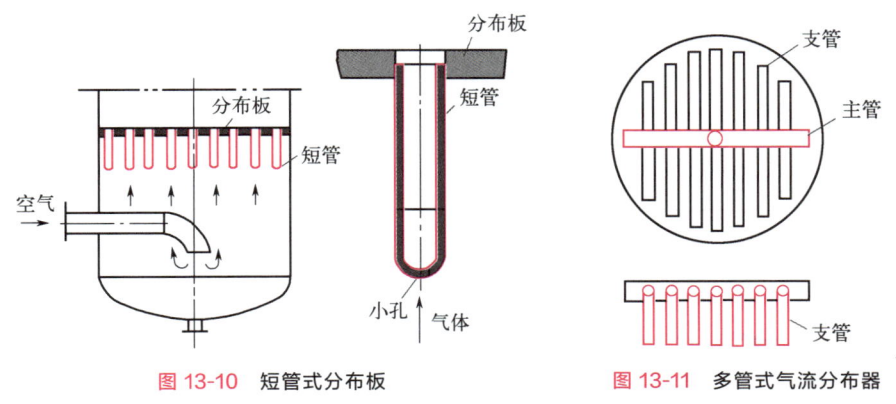

图 13-10　短管式分布板　　　　图 13-11　多管式气流分布器

气体分布板压降，gas distribution plate pressure drop，流体通过分布板的压降可用床内表观速度的速度头倍数来表示。

$$\Delta p_{D} = 9.807 C_{D} \frac{u^2 \rho_{f}}{2\varphi^2 g} \tag{13-14}$$

式中，Δp_D 为分布板压降，Pa；φ 为开孔率；C_D 为阻力系数，其值在 1.5～2.5 之间，对于锥帽侧缝式分布板取 2.0。

两相模型，two-phase model，如图 13-12 所示，基本思想是把流化床分成气泡相和乳化相，分别研究这两个相中的流动和传递规律，以及流体与颗粒在相间的交换。对于气、乳两相的流动模式则一般认为气相为置换流，而对乳化相则有种种不同的处理，如置换流、全混流、部分返混、环流或对其流动模式不加考虑等。

鼓泡床模型，bubbling bed model，如图 13-13 所示，有下列基本假设：①床层分为气泡区、泡晕及乳化相三个区域，在这些相间产生气体交换，这些气体交换过程是串联的；②乳化相处于临界流化状态，超过起始流化速度所需要的那部分气量以气泡的形式通过床层；③气泡的长大与合并主要发生在分布板附近的区域，因而假设在整个床层内气泡的大小是均匀的，认为气泡尺寸是决定床内情况的一个关键因素，这个气泡尺寸不一定就是实际的尺寸，因而称它为气泡有效直径；④只要气体流速大于起始流化速度的两倍，即 $u > 2u_{mf}$，床层鼓泡剧烈的条件便可满足，气泡内基本上不含固体颗粒；⑤乳化相中的气体可能向上流动，也可能向下流动，当 $u/u_{mf} > 6～11$ 时，乳化相中的气体从上流转为下流，虽然流向有所不同，但这部分的气量与气泡相相比甚小，对转化率的影响可忽略，此时离开床层的气体组成等于床层顶部处的气体组成，这样不必考虑乳化相中的情况，只需计算气泡中的气体组成便可计算反应的转化率。

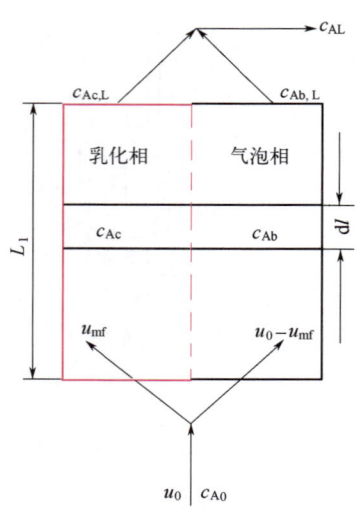

图 13-12　两相模型示意图

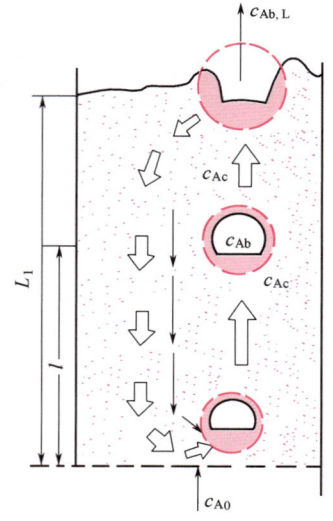

图 13-13　鼓泡床模型示意图

互动练习

13-11　What are the two factors affecting the mass transfer rates in fluidized bed reactors?

A）Concentration gradients and flow conditions

B）Particle size and shape

C）Gas velocity and temperature

D）Particle density and flow rate

13-12　Which modes of heat transfer are involved in fluidized bed reactors?

A）Conduction and convection

B）Convection and radiation

C）Conduction and radiation

D）Convection，conduction，and radiation

13-13　How can the heat transfer rates in fluidized bed reactors be enhanced?

A）By decreasing the particle size distribution

B）By increasing the bed temperature

C）By decreasing the gas velocity

D）By optimizing the reactor design and adjusting the fluidization parameters

13-14　What are the factors to consider in determining the diameter and height of a fluidized bed reactor?

A）Particle density and shape

B）Gas velocity and temperature

C）Bed diameter and bed height

D）Reactor design and operation

13-15　Which mathematical models are commonly used in fluidized bed reactor design?

A）One-phase model and fluidized bed model

B）Two-phase model and bubble column model

C）Three-phase model and moving bed model

D）Two-phase model and fluidized bed model

任务14

Operation and Control of Fixed Bed Reactors
固定床反应器操作与控制

任务要点

本任务讲解固定床反应器操作中的关键控制点，涵盖温度控制、流体分布及压降等内容。读者将学习如何优化设备运行条件，提高反应效率，并能快速诊断和解决运行中的常见故障。

学习目标

知识目标

（1）了解固定床反应器运行过程中常见的操作问题及其原因。

（2）掌握固定床反应器操作参数的控制方法。

技能目标

（1）能优化固定床反应器的运行条件。

（2）能制定有效的操作策略以确保生产连续性和安全性。

价值目标

（1）强调生产过程中固定床反应器安全操作的重要性。

（2）培养在固定床反应器运行中解决问题的能力。

14.1 Use Process of Industrial Catalyst
工业催化剂使用

Catalysts are essential in many industrial processes，and proper handling and use are crucial for achieving optimal performance.

Transport，storage，and filling of catalysts require special care to avoid contamination（污染）and ensure consistency（前后一致性）.

Catalysts must be carefully heated and reduced to activate them before use.

During operation，the catalyst must be periodically deactivated （周期性地钝化），or "poisoned"，to avoid unwanted side reactions or product degradation （降解）.

Finally，the catalyst may need to be regenerated （再生） or replaced when it becomes deactivated （失去活性） or worn out （废旧）.

Understanding the characteristics of the catalyst and the reaction environment is essential for effective catalyst use. Careful monitoring and adjustment of operating conditions can maximize the catalyst's efficiency and lifespan （有效期）. Proper disposal （妥善丢弃） of spent catalysts is also critical to minimize environmental impact. The safe handling and use of catalysts are crucial to achieving efficient and sustainable industrial processes.

技术理论

催化剂通常是装桶供应，有金属桶或纤维板桶包装。催化剂的贮藏要求防潮、防污染。催化剂的装填是非常重要的工作，装填好坏对催化剂床层气流的均匀分布（降低床层的阻力有效发挥催化剂的效能）有重要作用。图14-1即为搬运催化剂桶的装置。

催化剂的升温与还原是投入使用前的最后一道工序，也是催化剂形成活性结构的过程。

催化剂的失活包括中毒、积碳、烧结、挥发与剥落，再生方法包括蒸汽处理、空气处理、通入氢气或不含毒物的还原性气体和用酸或碱溶液处理。

图14-1 搬运催化剂桶的装置

催化剂的卸出应做好充分的准备工作，制定出详细的停工卸出方案。除了包括正常的降温、钝化内容外，还要安排废催化剂的取样工作，以便收集资料，帮助分析失活原因，同时安排好物资供应工作。

关键词详解

纤维板，fiberboard，纤维板是一种由木材和其他植物纤维经过破碎、蒸煮、剥离、成纤维和混合等工艺制作而成的板材。在催化剂的使用中，纤维板可以作为催化剂的载体，提供催化反应所需的表面积和承载力。

催化剂装填，catalyst loading，指将催化剂颗粒或成型块体填充到反应器中的操作过程。催化剂装填的目的是将催化剂均匀分布在反应器中，以保证催化反应的高效进行。图 14-2 与图 14-3 即展示了装填催化剂的装置和料斗。

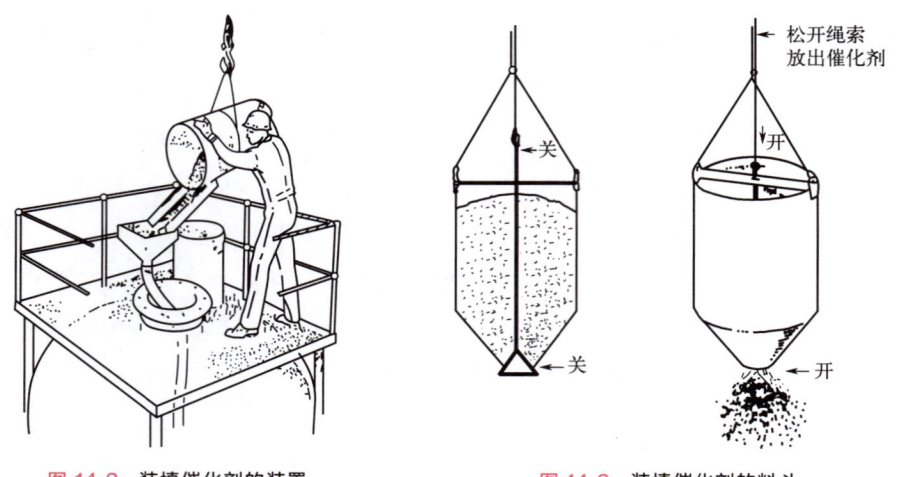

图 14-2 装填催化剂的装置   图 14-3 装填催化剂的料斗

催化剂升温与还原，catalyst heating and reduction，在催化反应开始之前，需要将催化剂进行升温处理，以达到催化反应所需的活化温度。还原是指通过加热或加氢等方式，将催化剂上的氧化物还原成金属或金属间化合物，以恢复催化剂活性。

催化剂失活，catalyst deactivation，催化剂失活是指由于催化剂受到污染、烧结、磨损等原因导致其催化活性下降或完全丧失的过程。失活的原因是各种各样的，主要是中毒、积炭、烧结、挥发与剥落等，如图 14-4 所示。

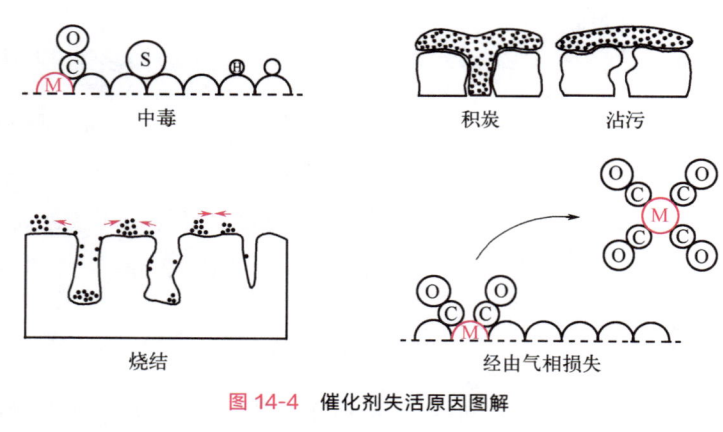

图 14-4 催化剂失活原因图解

M—金属

中毒，poison，催化剂中毒是指催化剂表面被吸附或吸附生成物，导致活性中心被占据或毒化，从而降低催化剂的活性和选择性。催化剂的毒物通常可分为化学型毒物和选择型毒物两大类。

积炭，carbon deposition，当催化剂与反应物接触时，某些碳氢化合物可能会在催化剂表面发生部分氧化和聚合反应，形成碳的沉积物，称为催化剂积炭。发生积炭的原因很多，通常是催化剂导热性能不好或孔隙过细时容易发生。积炭过程是催化系统中的分子经脱氢-聚合而形成难挥发性高聚物，它们还可以进一步脱氢而形成含氢量很低的类焦物质，所以积炭常称被为结焦。如图14-5所示，丁烷在铝-铬催化剂上脱氢时，结焦相当激烈。

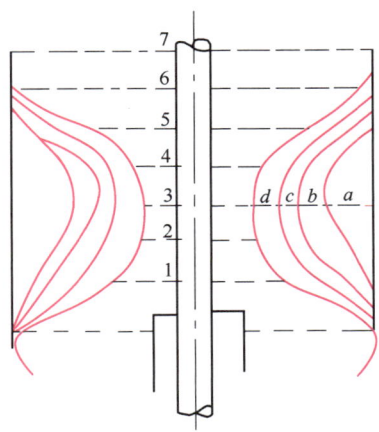

图 14-5 丁烷脱氢反应器中结焦的情况

a—最初结焦区；b～d—后来结焦区；1～7—反应器挡板

烧结，sinter，催化剂烧结是指催化剂颗粒在高温下相互接触和融合，形成较大的团聚体或孔隙闭塞，导致催化剂活性表面积减少、反应活性降低的现象。

催化剂再生，catalyst regeneration，催化剂再生是指对失活或毒化的催化剂进行修复和恢复活性的过程。常见的催化剂再生方法包括热氧化、焙烧、洗涤、还原等处理。

互动练习

14-1 What is the purpose of activating a catalyst before using it in a reaction?

A）To remove impurities from the catalyst

B）To increase the surface area of the catalyst

C）To increase the reactivity of the catalyst

D）To decrease the surface area of the catalyst

14-2 What is the process of passivation of a catalyst?

A）Removing the impurities from the catalyst

B）Increasing the surface area of the catalyst

C) Reducing the reactivity of the catalyst

D) Increasing the reactivity of the catalyst

14-3　What are the measures taken to prevent catalyst deactivation?

A) Regular cleaning of the reactor

B) Controlling the reaction conditions

C) Using a higher concentration of reactants

D) Increasing the reactor temperature

14-4　What is the purpose of catalyst regeneration?

A) To remove impurities from the catalyst

B) To restore the catalytic activity of the catalyst

C) To increase the surface area of the catalyst

D) To decrease the surface area of the catalyst

14-5　What is the purpose of catalyst deactivation?

A) To increase the catalytic activity of the catalyst

B) To reduce the surface area of the catalyst

C) To restore the catalytic activity of the catalyst

D) To reduce the reactivity of the catalyst

14.2　Operation and Control of Fixed Bed Reactors 固定床反应器操作与控制要点

Fixed bed reactors are widely used in the chemical industry for various reactions. To ensure efficient and safe operation of these reactors, several key factors must be considered.

Temperature control is a critical factor in maintaining the reaction rate and selectivity. A proper temperature profile（温度分布）must be established and maintained throughout the reactor bed.

Pressure control is also important to prevent pressure build-up（压力积累）, which can cause safety concerns.

The hydrogen to oil ratio should be carefully controlled to ensure the proper amount of hydrogen is available for the reaction.

The gas velocity should be maintained within a specific range to prevent excessive pressure drop（压降过大）and catalyst attrition（催化剂磨损）.

In addition, catalyst regeneration is a necessary step in maintaining the activity of the catalyst. The proper operating conditions, such as temperature and flow rate, must be established for efficient catalyst regeneration.

Proper control and operation of fixed bed reactors are crucial for achieving the desired reaction results and ensuring safe operation of the reactor.

技术理论

以加氢裂化反应器为例，固定床反应器的操作与控制要点，主要包括温度调节、压力调节、氢油比控制、空速操作原则和催化剂器内再生操作。

关键词详解

温度调节，temperature regulation，需注意：控制反应器入口温度，控制反应床层间的急冷氢量，原料组成的变化会引起温度的变化，反应器初期与末期的温度变化，反应温度的限制。

压力调节，pressure regulation，需注意：氢气压缩机的压力调节，反应温度的影响，原料变化的影响。

氢油比控制，hydrogen-oil ratio control，氢油比的大小或反应物循环量大小直接关系到氢分压和油品的停留时间，还影响油品的汽化率。

空速操作原则，airspeed operation principles，在操作过程中，需要进行提温提空速时，应"先提空速后提温"，而降空速降温时则"先降温后降空速"。

催化剂器内再生操作，regeneration operation inside the catalyst vessel，需注意：再生前的预处理（先降温，停止进料，用惰性气体吹扫系统），再生的进行（注意控制一定的升温速率），再生的结束（小心观察床层内各点温度）。

互动练习

14-6　What are the key points for operating and controlling a fixed bed reactor?

A）Temperature control，pressure control，hydrogen-oil ratio control，airspeed operation principles，catalyst regeneration operation

B）Catalyst storage and transportation，temperature control，pressure control，hydrogen-oil ratio control，airspeed operation principles

C）Catalyst activation and reduction，pressure control，hydrogen-oil ratio control，airspeed operation principles，catalyst regeneration operation

D）Temperature control，pressure control，hydrogen-oil ratio control，airspeed operation principles，catalyst deactivation

14-7　What is the principle of hydrogen-oil ratio control in fixed bed reactors?

A）Adjusting the ratio of hydrogen to oil in the reactor to ensure a high conversion rate of the feedstock

B）Maintaining a constant ratio of hydrogen to oil in the reactor to prevent explosions

C）Increasing the ratio of hydrogen to oil in the reactor to reduce carbon deposition on the catalyst

D）Decreasing the ratio of hydrogen to oil in the reactor to improve the selectivity of the reaction

14-8　What is the principle of airspeed operation in fixed bed reactors?

A）Maintaining a constant flow rate of air in the reactor to promote catalyst regeneration

B）Adjusting the flow rate of air in the reactor to maintain a constant temperature

C）Controlling the flow rate of air in the reactor to prevent excessive pressure drop

D）Increasing the flow rate of air in the reactor to improve the heat transfer efficiency

14-9　How can catalyst deactivation be prevented in fixed bed reactors?

A）By adjusting the temperature and pressure of the reactor to optimal conditions

B）By controlling the hydrogen-oil ratio to prevent excessive carbon deposition on the catalyst

C）By using a catalyst with high resistance to deactivation

D）By periodically regenerating the catalyst in the reactor

14-10　What is the purpose of catalyst regeneration in fixed bed reactors?

A）To remove impurities and contaminants from the catalyst

B）To improve the activity and selectivity of the catalyst

C）To prevent excessive carbon deposition on the catalyst

D）To reduce the pressure drop across the catalyst bed

14.3 Operation and Control of Ethylbenzene Dehydrogenation Reactors 乙苯脱氢反应器操作与控制

Ethylbenzene dehydrogenation（乙苯脱氢）is a common process used in the production of styrene. The process involves the dehydrogenation of ethylbenzene to form styrene and hydrogen gas over a fixed-bed catalyst.

To ensure safe and efficient operation of the reactor，several key points must be considered. Firstly，temperature control is crucial to maintaining the desired reaction rate and preventing catalyst deactivation. Secondly，pressure control is important to prevent overpressure（超压）situations that can damage the reactor or cause safety hazards. Thirdly，the hydrogen-oil ratio must be carefully controlled to maintain optimal reaction conditions. Additionally，proper operation principles for gas flow rate should be followed to avoid uneven catalyst bed distribution（不均匀催化剂床层分布）or channeling（沟流）. Finally，abnormal situations such as steam outage（蒸汽中断），power outages（停电），and other disturbances（故障），must be considered and addressed appropriately（适当考虑与处理）.

During normal operation，the reactor should be carefully monitored，and any abnormalities addressed immediately. Prior to startup（开车），the reactor should be prepared，and the catalyst bed checked for any defects or abnormalities. During normal operation，the reactor should be carefully controlled to maintain the desired temperature，pressure，and gas flow rate. When stopping the reactor，proper procedures must be followed to ensure safe and efficient shutdown（停车）.

In the event of an abnormal situation，the reactor must be immediately shut down and appropriate measures taken to address the issue. Common abnormal situations include steam or power outages，catalyst deactivation，and channeling or uneven catalyst bed distribution. In these situations，the reactor must be carefully inspected，and appropriate corrective actions taken.

技术理论

乙苯绝热脱氢生产工艺流程如图 14-6 所示，向来自乙苯蒸发器的乙苯蒸气中加入占总配料 1.5% 的水蒸气进行稀释，利用脱氢后的物料进行加热，使温度达到 150℃，再进入乙苯过热器过热，进一步与脱氢气换热至 500℃ 后进入进料混合器，与来自水蒸气过热炉的温度为 770℃ 的过热水蒸气进行混合，控制温度在 630℃ 左右，然后进入乙苯脱氢反应器，反应器为单壳圆筒双段绝热式，经一段反应后温度降为 580℃，再与来自水蒸气过热炉的过热水蒸气混合，继续进行二段反应，反应后的气体温度降为 590℃，与乙苯过热器、废热锅炉、乙苯蒸发器进行热交换后，温度降为 137℃ 左右，然后进入冷凝器，液相进入油水分离器，油层送入脱氢液储罐供精馏使用。

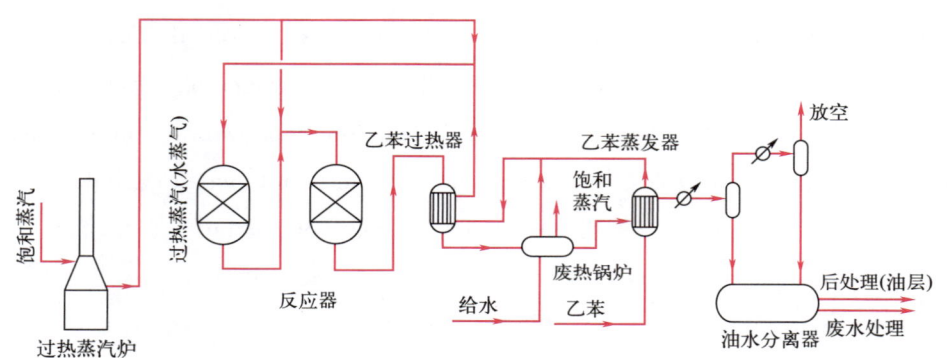

图 14-6 乙苯绝热脱氢工艺流程示意图

关键词详解

开车，start up，本工艺开车之前进行开车前准备工作，经过燃料管道吹扫、炉膛吹扫、点火后，进行化工投料操作，观察反应釜内温度和压力上升情况，控制适当的升温速度，逐渐使反应温度、压力等工艺指标达到正常值并进行稳定运行。

正常运行，normal operation，本工艺的正常运行包括：

① 本岗位所有温度、压力、流量、液位均应每小时如实记录一次，数据要正确无误，字迹端正，不得涂改。

② 对本岗位所属管道、设备每小时应检查一次，发现异常及时汇报，并做好记录。

③ 经常观察加热炉燃烧情况，调节喷嘴火焰，稳定炉顶温度。

④ 需要增加负荷时，先加水蒸气负荷，后加乙苯负荷；要减少负荷时，先减乙苯负荷，后减水蒸气负荷。

停车，shut down，本工艺停车操作其过程如下：

① 接到停车通知后，逐步减少乙苯进料流量，以 10℃/h 速率降低炉顶温度至 800℃后恒温。

② 在 800℃恒温下，仍按一定的速率减少乙苯进料量，直至切断乙苯。

③ 800℃恒温结束后，以 15℃/h 速率降低炉顶温度至 750℃，关小烟囱挡板角度。

④ 750℃恒温 1h，逐步减少水蒸气进入量，再关小烟囱挡板角度，以减少空气进入量，关闭盐水阀。

⑤ 以 15℃/h 速率降低炉顶温度至 500℃，减少水蒸气进入量。

⑥ 500℃恒温 17h，恒温过程中，第三小时开始进一步减少水蒸气进入量，交替切换动力空气，控制动力空气的流量。

⑦ 恒温结束后，以 15℃/h 速率降低炉顶温度至 150℃，继续以一定流量通动力空气。

⑧ 150℃恒温 2h，关小烟囱挡板角度。

⑨ 恒温结束切断动力空气阀，关小烟囱挡板角度。并以 20℃/h 速率降低炉顶温度至熄火，然后自然降温。

⑩ 切断循环上水，排净存水，必要时要加盲板。

互动练习

14-11　What is the purpose of the preparation work before starting up the reactor?

A）To ensure safety

B）To increase production

C）To decrease production costs

D）To save time

14-12　What should be done before shutting down the reactor?

A）Stop the flow of feedstock

B）Turn off the reactor's heaters

C）Purge the system with inert gas

D）All of the above

14-13　What should be done if there is a sudden power outage?

A）Open the reactor's relief valves

B）Turn off the reactor's heaters

C）Contact the plant's control room

D）Both A and C

14-14　What is one common abnormality in an adiabatic fixed bed reactor?

A) A decrease in temperature

B) An increase in temperature

C) A decrease in pressure

D) An increase in pressure

14-15　What is the purpose of the hydrogen-oil ratio control in the reactor?

A) To increase the reactor's efficiency

B) To decrease the reactor's efficiency

C) To maintain a constant temperature

D) To reduce the production of unwanted byproducts

任务15

Operation and Control of Fluidized Bed Reactors
流化床反应器操作与控制

任务要点

本任务重点介绍流化床反应器的操作与控制技术，涵盖气速调节、温度分布及流化稳定性等内容。读者将学习如何通过优化运行条件提高设备效率，并确保设备长期稳定运行。

学习目标

知识目标

（1）了解流化床反应器操作中的关键控制变量及其影响。

（2）掌握流化床反应器在操作中的动态行为和优化策略。

技能目标

（1）能识别和解决流化床反应器运行中的操作问题。

（2）能调节关键参数以提高产品质量和反应效率。

价值目标

（1）注重流化床反应器的安全运行。

（2）在流化床反应器操作中贯彻节能减排理念。

15.1 Operation and Control of Fluidized Bed Reactors 流化床反应器操作与控制要点

Fluidized bed reactors are widely used in the chemical industry due to their excellent mass and heat transfer properties. Proper operation and control of these reactors are essential for ensuring safe and efficient operation. There are several key points to consider when operating and controlling a fluidized bed reactor.

Firstly，the control of particle size and composition is critical for maintaining consistent reactor performance. Particle size and composition affect the reactor's fluidization behavior（流化行为）, which in turn affects the heat and mass transfer rates.

Secondly，pressure measurement and control（压力测量与控制）are essential for maintaining stable reactor operation. Pressure measurement can detect abnormal conditions，and pressure control can help prevent catastrophic failure（灾难性故障）.

Thirdly，temperature measurement and control are necessary to maintain a stable reaction rate and prevent hot spots from developing. Accurate temperature control is critical to ensuring product quality and avoiding undesirable by-products（避免不必要的副产物）.

Fourthly，flow control（流量控制）is essential to maintain consistent product quality and prevent gas channeling，which can lead to uneven flow distribution and localized hot spots（局部热点）.

Finally，proper startup and shutdown procedures must be followed，and safety protocols must be established and followed to prevent accidents. The maintenance and inspection of the reactor and associated equipment are also critical for safe and efficient operation.

In summary，careful control of particle size，pressure，temperature，and flow rate，as well as proper startup and shutdown procedures and maintenance，are essential for the safe and efficient operation of a fluidized bed reactor.

技术理论

对于一般的工业流化床反应器，需要控制和测量的参数主要有颗粒粒度、颗粒组成、床层压力和温度、流量等。这些参数的控制除了受所进行的化学反应的限制外，还要受到流态化要求的影响。实际操作中是通过安装在反应器上的各种测量仪表了解流化床中的各项指标，以便采取正确的控制步骤，达到反应器的正常工作。

关键词详解

颗粒粒度和组成的测量与控制，measurement and control of particle size and composition，颗粒粒度和组成对流态化质量和化学反应转化率有重要影响。在氨氧化制丙烯腈的反应器内，采用的催化剂粒度和组成中，为了保持小于 $44\mu m$ 的"关键组分"（即对流态化质量起关键作用的较小粒度的颗粒）粒子在 20%～40% 之间，通常采用激光粒度分析仪、显微镜检查、化学分析等手段进行测量，并通过安装在反应器上的"造粉器"实时调整。当发现床层内小于 $44\mu m$ 的粒子小于 12% 时，就启动造粉器。系统正常运转中，从床层取固体颗粒样品，虽然简单，但又要特别注意并妥善处理。图 15-1 所示就是一种常用的流化床用的固体颗粒取样器。

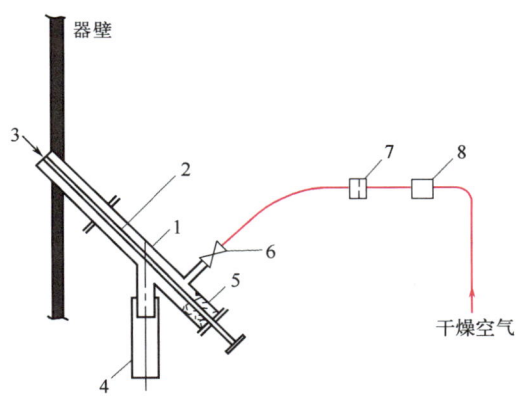

图 15-1　流化床用固体颗粒取样器

1—取样器本体，$\phi(32\sim38)$mm；2—拉杆；3—锥形活动堵头；4—手动装卸取样杯，$\phi38$mm×150mm；

5—气密填料；6—针形阀；7—节流小孔板，孔 $\phi(0.6\sim1.0)$mm；8—逆止阀

压力的测量与控制，measurement and control of pressure，压力和压降的测量，是了解流化床各部位是否正常工作较直观的方法。对于实验室规模的装置，U 形管压力计是常用的测压装置，通常压力计的插口需配置过滤器，以防止粉尘进入 U 形管。工业装置上常采用带吹扫气的金属管做测压管。测压管线的典型安装如图 15-2 所示。

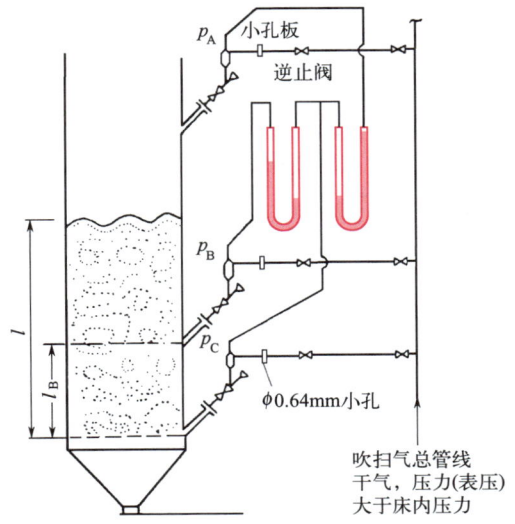

图 15-2　流化床差压计安装管线示意图

温度的测量与控制，measurement and control of temperature，流化床催化反应器的温度控制取决于化学反应的最优反应温度的要求。一般要求床内温度分布均匀，符合工艺要求的温度范围。通过温度测量可以发现过高温度区，进一步判断产生的原因是存在死区，还是反应过于剧烈，或者是换热设备发生故障。通常由于存在死区造成的高温，可及时调整气体流量来改变流化状态，从而消除死区。如果是因为反应过于激烈，可以通过调节反应物流量或配比加以改变。换热器是保证稳定反应温度的重要装置，正常情况下通过调节加热或制冷剂的流量就能保证工艺对温度的要求。最常用的温度测量办法是采用标准的热敏元件，如适应各种温度范围测量的热电偶。

流量的测量与控制，measurement and control of flow rate，气体的流量在流化床反应器中是一个非常重要的控制参数，它不仅影响着反应过程，而且关系到流化床的流化效果。所以，作为既是反应物又是流化介质的气体，其流量必须要在保证最优流化状态下有较高的反应转化率。一般原则是达到最优流化状态所需的气速后，应在不超过工艺要求的最高或最低反应温度的前提下尽可能提高气体流量，以获得最高的生产能力。气体流量的测量一般采用孔板流量计，要求被测的气体是清洁的。

开停车及事故防止，start up，shut down，and accident prevention，由粗颗粒形成的流化床反应器，开车启动操作一般不存在问题。而细颗粒流化床，特别是采用旋风分离器的情况下，开车启动操作需按一定的要求来进行。这是因为细颗粒在常温下容易团聚。

正常的停车操作对保证生产安全，减少对催化剂和设备的损害，为开车创造有利条件等都是非常重要的。不论是对固相加工还是气相加工，正常停车的顺序都是首先切断热源（对于放热反应过程，则是停止送料），随后降温。

互动练习

15-1　Which of the following is important for controlling particle size in a fluidized bed reactor?

　　A）Reactor diameter

　　B）Bed height

　　C）Gas distribution plate

　　D）Temperature control

15-2　What is the primary method for controlling pressure in a fluidized bed reactor?

　　A）Adjusting the reactor diameter

B）Controlling gas flow rate

C）Controlling particle size

D）Varying the bed height

15-3　How is temperature controlled in a fluidized bed reactor?

A）By adjusting the gas flow rate

B）By varying the bed height

C）By using a heat exchanger

D）By adjusting the temperature of the inlet gas

15-4　Which of the following is an important consideration for flow rate control in a fluidized bed reactor?

A）Reactor diameter

B）Particle size distribution

C）Bed height

D）Gas distribution plate design

15-5　What is an important measure for preventing accidents in fluidized bed reactor operations?

A）Ensuring consistent particle size

B）Maintaining a constant reactor temperature

C）Regularly checking and maintaining gas distribution plates

D）Properly following start up and shutdown procedures

15.2 Operation and Control of Propane-Based Polypropylene Fluidized Bed Reactor 本体聚合流化床反应器的操作与控制

The HIMONT polypropylene（聚丙烯）process uses a fluidized bed reactor for non-catalytic polymerization（非催化聚合）. In this process, propylene is polymerized into polypropylene, a widely used thermoplastic（热塑性）material. The reactor operates in a continuous mode, allowing for efficient production with high yields.

The operation and control of the reactor involves several steps. Before starting the reactor, proper checks and preparations are made to ensure the safety and efficiency of the process. During the startup, the reactor is gradually heated and the feed rate of propylene is gradually increased until a steady-state（稳定状态）condition is achieved. Once the steady-state condition is reached，the reactor is operated at the desired temperature，pressure，and propylene flow rate.

To stop the reactor，the propylene feed rate is reduced gradually，and the reactor is cooled down slowly. In case of an emergency，an emergency stop （紧急停车）can be initiated to quickly stop the reactor.

Common abnormal phenomena that can occur during operation of the fluidized bed reactor include agglomeration，fouling，and sheeting（表面集片）. Agglomeration occurs when particles stick together and form large lumps（大块），which can lead to reactor blockages（堵塞）. Fouling occurs when the reactor walls become coated with polymer，which can affect the heat transfer efficiency. Sheeting occurs when the polymer forms a thin film on the reactor surface，which can also affect heat transfer and cause fouling.

To prevent these issues，regular maintenance and cleaning of the reactor are necessary. Proper operation and control of the reactor can also minimize the occurrence of abnormal phenomena.

技术理论

生产工艺流程如图 15-3 所示。具有剩余活性的干均聚物（聚丙烯）在压差

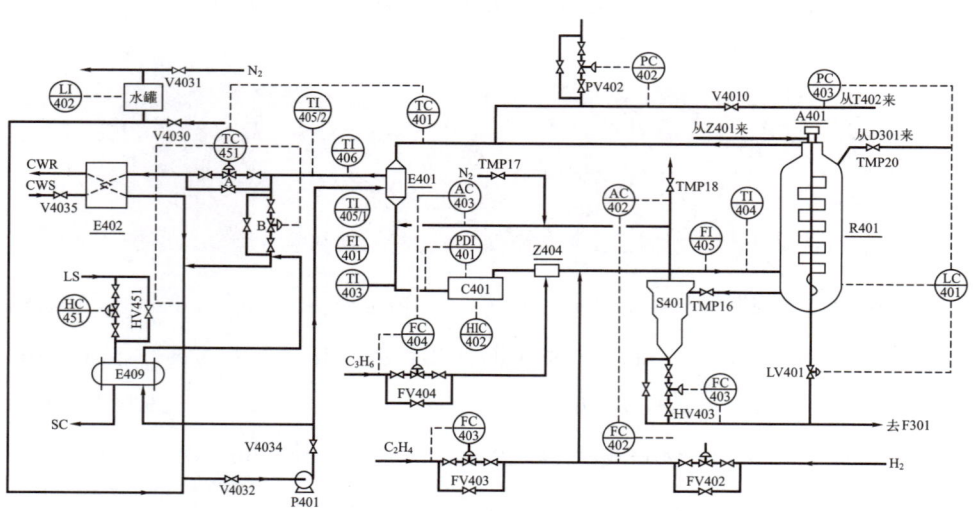

图 15-3　高抗冲击共聚物生产工艺流程图

A401—刮刀；C401—循环压缩机；D301—闪蒸罐；E401、E402—冷却器；

E409—夹套水加热器；F301—火炬；P401—开车加热泵；R401—反应器；

S401—旋风分离器；T402—乙烯汽提塔；Z401—过滤器

作用下自闪蒸罐流到气相共聚反应器。在气体分析仪的控制下，氢气被加到乙烯进料管道中，以改进聚合物的本征黏度，满足加工需要。

聚合物从顶部进入流化床反应器，落在流化床的床层上。流化气体（反应单体）通过一个特殊设计的栅板进入反应器。由反应器底部出口管路上的调节阀来维持聚合物的料位。聚合物料位决定了停留时间，从而决定了聚合反应的程度，为了避免过度聚合的鳞片状产物堆积在反应器壁上，反应器内配置一转速较慢的刮刀，以使反应器壁保持干净。栅板下部夹带的聚合物细末用一台小型旋风分离器除去，并送到下游的袋式过滤器中。

所有未反应的单体循环返回到流化压缩机的吸入口。来自乙烯气体提升塔顶部的回收气相与气相反应器出口的循环单体汇合，而补充的氢气、乙烯和丙烯加入到压缩机排出口循环气体用工业色谱仪进行分析，调节氢气和丙烯的补充量。然后调节补充的丙烯进料量，以保证反应器的进料气体满足工艺要求的组成。

关键词详解

开车，start up，主要包括：进行反应进料；准备接收来自 D301 的均聚物；共聚反应物的开车。

正常运行，normal operation，注意控制反应器的液位和反应器压力和气相组成。

停车，shut down，主要包括：降低反应器料位；关闭乙烯进料，保压；关闭丙烯及氢气进料；氮气吹扫。

互动练习

15-6　What is the purpose of the HIMONT polypropylene process?

A）To produce propylene

B）To produce thermoplastic material

C）To produce polyethylene

D）To produce catalyst for polymerization

15-7　How does the reactor operate in the HIMONT polypropylene process?

A）It operates in a batch mode

B）It operates in a semi-batch mode

C）It operates in a continuous mode

D）It operates in a discontinuous mode

15-8　　What is the procedure for stopping the reactor in the HIMONT polypropylene process?

A）The reactor is cooled down slowly and the propylene feed rate is reduced gradually

B）The reactor is cooled down quickly and the propylene feed rate is reduced suddenly

C）The reactor is heated up and the propylene feed rate is increased suddenly

D）The reactor is heated up and the propylene feed rate is increased gradually

15-9　　What are some common abnormal phenomena that can occur during operation of the fluidized bed reactor?

A）Agglomeration，fouling，and sheeting

B）Dehydration，hydrolysis，and oxidation

C）Polymerization，polymer degradation，and crosslinking

D）Cracking，coking，and carbonization

15-10　　What is necessary to prevent abnormal phenomena in the HIMONT polypropylene process?

A）Regular maintenance and cleaning of the reactor

B）Increasing the propylene feed rate

C）Reducing the reactor temperature

D）Decreasing the propylene flow rate

15.3　Common Abnormal Phenomena and Handling Methods in Fluidized Bed Reactors
流化床反应器中常见的异常现象及处理方法

Fluidized bed reactors are widely used in chemical and petrochemical industries due to their high heat and mass transfer rates. However，several abnormal phenomena can occur during the operation of fluidized bed reactors，which can affect the reactor's performance and safety.

The first common anomaly（异常）is channeling，which refers to the formation of a localized channel（局部沟流）or pathway in the reactor bed where the fluidization is incomplete. Channeling can result in inefficient use（低效使用）of catalyst or reactants，reduced conversion rate，and hot spots in the bed. To overcome channeling，several techniques can be employed，such as introducing a pulsating gas flow（脉动气流）or redistributing（再分布）the catalyst particles.

The second anomaly is the formation of large bubbles，which can cause uneven distribution of the fluidized material and affect the contact between the reactants and catalyst. Large bubbles can also lead to the accumulation of solids，resulting in blockage of the reactor. To mitigate（减少）large bubbles，reducing the superficial gas velocity（表面气速），increasing the particle size，and adjusting the fluidization velocity can be considered.

The third anomaly is the phenomenon of bed expansion or bubbling（床层膨胀或起泡），known as the "churn-turbulent regime（湍流状态）". This can lead to a decrease in reaction efficiency and affect the catalyst's stability. To control bubbling，adjusting the fluidization velocity，particle size，and bed height can be considered.

技术理论

流化床反应器中常见的异常现象包括沟流现象、大气泡现象和腾涌现象。

关键词详解

沟流现象，channeling phenomenon，沟流现象的特征是气体通过床层时形成短路，如图 15-4 所示。沟流有两种情况：（a）图所示的贯穿沟流和（b）图所示的局部沟流。沟流现象发生时，大部分气体没有与固体颗粒很好接触就通过了床层，这在催化反应时会引起催化反应的转化率降低。由于部分颗粒没有流化或流化不好，造成床层温度不均匀，从而引起催化剂的烧结，降低催化剂的寿命和效率。因为沟流时部分床层为死床，不悬浮在气流中，故在 Δp-u 图上反映出 Δp 始终低于理论值 W/A，如图 15-5 所示。

沟流现象产生的原因主要与颗粒特性和气体分布板的结构有关。下列情况容易产生沟流：颗粒的粒度很细（粒径小于 $40\mu m$）、密度大且气速很低时；潮湿的物料和易于黏结的物料；气体分布板设计不好，气体分布不均，如孔太少或各个风帽阻力大小差别较大。

要消除沟流，应对物料预先进行干燥并适当加大气速，另外分布板的合理设计也是十分重要的。还应注意风帽的制造、加工和安装，以免通过风帽的流体阻力相差过大而造成布气不均。

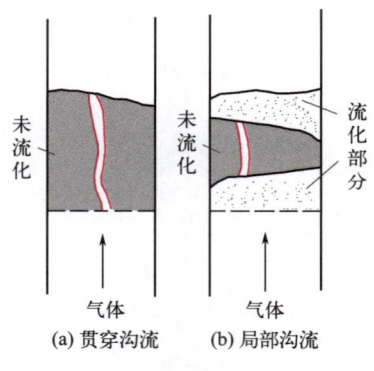

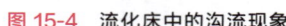

(a) 贯穿沟流　(b) 局部沟流

图 15-4　流化床中的沟流现象

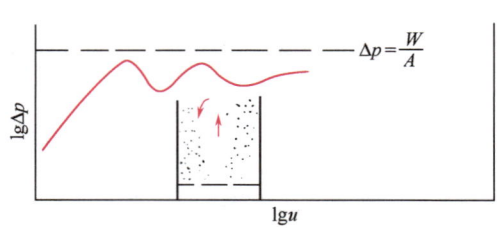

图 15-5　沟流时 Δp-u 的关系

大气泡现象，large bubble phenomenon，流化床中生成的气泡在上升过程中不断合并和长大，直到床面破裂是正常现象。但是，如果床层中大气泡很多，由于气泡不断搅动和破裂，床层波动大，操作不稳定，气固间接触不好，就会使气固反应效率降低，这是一种不正常现象，应力求避免。通常床层较高、气速较大时容易产生大气泡现象。在床层内加设内部构件可以避免产生大气泡，促使平稳流化。

腾涌现象，slugging phenomenon，所谓腾涌现象，就是在大气泡状态下继续增大气速，当气泡直径大到与床径相等时，就会将床层分为几段，变成一段气泡和一段颗粒的相互间隔状态，此时颗粒层被气泡像活塞一样向上推动，达到一定高度后气泡破裂，引起部分颗粒的分散下落。出现腾涌现象时，由于颗粒层与器壁的摩擦造成压降大于理论值，而气泡破裂时又低于理论值，即压降在理论值上下大幅度波动。腾涌发生时，床层的均匀性被破坏，使气固相的接触不良，严重影响产品的产量和质量，并且器壁磨损加剧，引起设备的振动。一般来说，床层越高、容器直径越小、颗粒越大、气速越高，越容易发生腾涌现象。在床层过高时，可以增设挡板以破坏气泡的长大，避免腾涌发生。

互动练习

15-11　What is the cause of channeling phenomenon in fluidized bed reactors?

A）Uneven gas flow rate

B）Agglomeration of particles

C）Wall effects

D）Poor design of the distributor plate

15-12　What is the solution for preventing bubbling fluidized bed reactors from large bubbles formation?

A）Increasing the bed height

B) Reducing the gas flow rate

C) Using smaller particles

D) Increasing the gas velocity

15-13   What is the main cause of the bubbling phenomenon in fluidized bed reactors?

A) Too high gas flow rate

B) Too low gas flow rate

C) Insufficient gas distribution

D) Excessive heat generation

15-14   What is the potential safety risk associated with the slugging phenomenon in fluidized bed reactors?

A) Explosions

B) Blockage of the reactor

C) Overheating

D) Corrosion

15-15   What is the recommended approach for handling the channelling phenomenon in fluidized bed reactors?

A) Increase the gas flow rate

B) Reduce the bed height

C) Adjust the design of the distributor plate

D) Decrease the particle size

项目三

Selection，Design，Operation and Control of Gas-Liquid Phase Reactors

气液相反应器选择、设计、操作与控制

项目三

Selection，Design，Operation and Control of Gas-Liquid Phase Reactors
气液相反应器选择、设计、操作与控制

任务16

Selection of Gas–Liquid Phase Reactors
气液相反应器选择

任务要点

本任务介绍气液相反应器的分类和适用工艺，读者将学习如何根据不同反应速率和传质要求选择合适的设备，并理解反应器选型对化工生产效率和经济性的影响。

学习目标

知识目标

（1）了解气液相反应器的分类及其适用工艺条件。

（2）掌握气液相反应器在选择时需考虑的因素，如反应速率和传质效率。

技能目标

（1）能根据不同化学反应的特性选择适合的气液相反应器。

（2）能评估反应器类型对生产成本和效率的影响。

价值目标

（1）提高反应器选择的科学性和经济性。

（2）培养对新型气液相反应器的探索和应用能力。

16.1 Classifican and Industrial Applications of Gas-Liquid Phase Reactors
气液相反应器分类和工业应用

Gas-liquid phase reactions（气液相反应）refer to chemical reactions that take place between a gas and a liquid phase. This type of reaction is commonly used in the chemical industry to produce a wide range of products，including pharmaceuticals，polymers，and food additives.

The main advantage of gas-liquid reactions is that they can be carried out under mild conditions，such as room temperature and atmospheric pressure. This makes them highly efficient and cost-effective compared to other types of reactions. Additionally，gas-liquid reactions are highly selective，meaning that they only react with specific molecules，which leads to high product yields（高收率）and reduced waste（低废弃物）.

There are several types of gas-liquid reactors that are commonly used in industrial applications. These include bubble column reactors（鼓泡塔反应器），bubble stirred tank reactors（鼓泡搅拌釜式反应器），and airlift reactors（气提反应器）. Bubble column reactors are known for their high gas-liquid contact area（气液接触面积），which leads to efficient gas-liquid mass transfer. Bubble stirred tank reactors are commonly used for reactions that require intense mixing，while airlift reactors are ideal for processes that require low shear rates（剪切速率）.

技术理论

气液相反应广泛地应用于加氢、磺化、卤化、氧化等化学加工过程。除此以外，气体产品的净化过程和废气及污水的处理过程以及好氧性微生物发酵过程均应用气液相反应过程。

如图 16-1 所示，气液相反应器按气液相接触形态可分为：①气体以气泡形态分散在液相中的鼓泡塔反应器、搅拌鼓泡釜式反应器和板式反应器；②液体以液滴状分散在气相中的喷雾、喷射和文氏反应器等高速湍动反应器；③液体以膜状运动与气相进行接触的填料塔反应器和降膜反应器等。

关键词详解

气液相反应，gas-liquid phase reaction，是指气体在液体中进行的化学反应。气体反应物可能是一种或多种；液体可能是反应物，或者只是催化剂的载体。

填料塔反应器，packed tower reactor，广泛应用于气体吸收的设备，也可用作气液相反应器。由于液体沿填料表面下流，在填料表面形成液膜而与气相接触进行反应，故液相主体量较少，适用于瞬间反应、快速和中速反应过程。例如，催化热碱吸收 CO_2、水吸收 NO_x 形成硝酸、水吸收 HCl 生成盐酸、水吸收 SO_3 生成硫酸等通常都使用填料塔反应器。

板式塔反应器，plate tower reactor，在板式塔反应器中液体是连续相，而

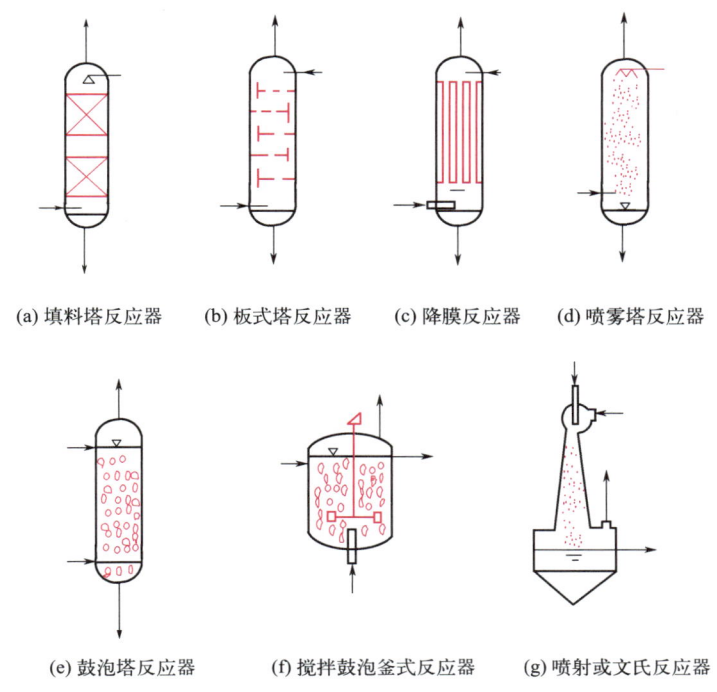

(a) 填料塔反应器　(b) 板式塔反应器　(c) 降膜反应器　(d) 喷雾塔反应器

(e) 鼓泡塔反应器　　(f) 搅拌鼓泡釜式反应器　　(g) 喷射或文氏反应器

图 16-1　气液相反应器的主要类型示意图

气体是分散相，借助于气相通过塔板分散成小气泡而与板上的液体相接触进行化学反应。板式塔反应器适用于快速及中速反应。

喷雾塔反应器，spray tower reactor，结构较为简单，液体以细小液滴的方式分散于气体中，气体为连续相，液体为分散相，具有相接触面积大和气相压降小等优点。适用于瞬间、界面和快速反应，也适用于生成固体的反应。

降膜反应器，falling film reactor，为膜式反应设备。通常借助管内的流动液膜进行气液反应，管外使用载热流体导入或导出反应热。降膜反应器可用于瞬间、界面和快速反应，它特别适用于较大热效应的气液反应过程。

搅拌鼓泡釜式反应器，stirred bubble tank reactor，是在鼓泡塔反应器的基础上加上机械搅拌以增大传质效率发展起来的。在机械搅拌的作用下，反应器内气体能较好地分散成细小的气泡，增大气液接触面积，但由于机械搅拌使反应器内液体流动接近全混流，同时能耗较高。釜式反应器适用于慢反应，尤其对高黏性的非牛顿型液体更为适用。

高速湍动反应器，high-speed turbulent reactor，喷射反应器、文氏反应器等属于高速湍动接触设备，它们适用于瞬间反应。此时，由于湍动的影响，加速了气膜传递过程的速率，因而获得很高的反应速率。

互动练习

16-1　What is a gas-liquid reaction?

A）A chemical reaction that takes place between a gas and a solid phase

B）A chemical reaction that takes place between a gas and a liquid phase

C）A chemical reaction that takes place between a liquid and a solid phase

D）A chemical reaction that takes place between two gas phases

16-2　What is the main advantage of gas-liquid reactions?

A）They can be carried out under high pressure

B）They can be carried out under extreme temperatures

C）They can be carried out under mild conditions

D）They can only react with specific molecules

16-3　What is the most used gas-liquid reactor for reactions that require intense mixing?

A）Bubble column reactor

B）Bubble stirred tank reactor

C）Airlift reactor

D）None of the above

16-4　What is the main reason for the high efficiency of gas-liquid reactions?

A）They require high temperatures

B）They require high pressures

C）They have high selectivity

D）They have low selectivity

16-5　What are the industrial applications of gas-liquid reactions?

A）Producing pharmaceuticals，polymers，and food additives

B）Producing electronic devices

C）Producing construction materials

D）Producing furniture

16.2 Structure of Bubble Column Reactors
鼓泡塔反应器结构

Bubble column reactors are widely used in the chemical industry to carry out gas-liquid reactions. These reactors are classified into several types based on their structure and mode of operation，including hollow type（空心式），multistage type（多段式），gas-lift type（气提式），and liquid jet type（液体喷射式）. Each type of reactor has unique features that make it suitable for specific applications.

The structure of a bubble column reactor consists of three main parts：the gas distributor（气体分布器）at the bottom，the column body（塔体），and the gas-liquid separator（气液分离器）at the top. The gas distributor is responsible for distributing the gas evenly throughout the reactor，while the column body is where the gas-liquid reaction takes place. The gas-liquid separator is used to separate the gas from the liquid phase.

The gas distributor is typically a porous plate（多孔板）or a sparger（喷淋器），which allows the gas to pass through and form bubbles in the liquid phase. The column body is usually cylindrical in shape and can be made of different materials such as glass，metal，or plastic. The diameter of the column body can range from a few centimeters to several meters，depending on the scale of the reaction. The gas-liquid separator is designed to ensure efficient separation of the gas phase from the liquid phase.

Overall，the structure of a bubble column reactor is critical to achieving efficient gas-liquid mass transfer and reaction. The design of the reactor should be optimized to ensure adequate gas-liquid contact（充分的气液接触），minimal pressure drop（最小化压降），and efficient separation of the gas phase from the liquid phase. By selecting the appropriate type of bubble column reactor and optimizing its structure，chemical engineers can carry out gas-liquid reactions more efficiently and economically.

技术理论

鼓泡塔反应器的基本组成部分主要有下述三部分：塔底部的气体分布器、塔筒体部分和塔顶部的气液分离器。

关键词详解

鼓泡塔反应器，bubble column reactor，鼓泡反应器是以液相为连续相，气相为分散相的气液反应器。液体分批加入，气体连续通入的称为半连续操作鼓泡塔。连续操作的鼓泡塔气体和液体连续加入，流动方向可以为向上并流或逆流。鼓泡塔多为空塔，一般在塔内设有挡板，以减少液体返混；为加强液体循环和传递反应热，可设外循环管和塔外换热器。图 16-2 所示为简单鼓泡塔反应器类型。图 16-3 所示为空心式鼓泡塔，这类反应器在化学工业上得到了广泛的应用，最适用于缓慢化学反应系统或伴有大量热效应的反应系统。若热效应较大时，可在塔内或塔外装备热交换单元，图 16-4 所示为具有塔内热交换单元的鼓泡塔。

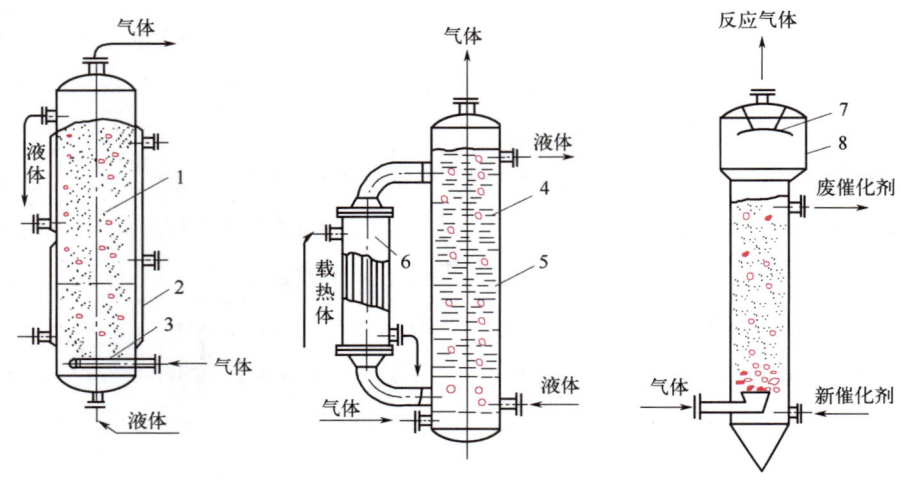

图 16-2 简单鼓泡塔反应器

1—塔体；2—夹套；3—气体分布器；4—塔体；5—挡板；6—塔外换热器；

7—液体捕集器；8—扩大段

动画

-鼓泡塔反应器原理图
-具有塔内热交换单元的鼓泡塔

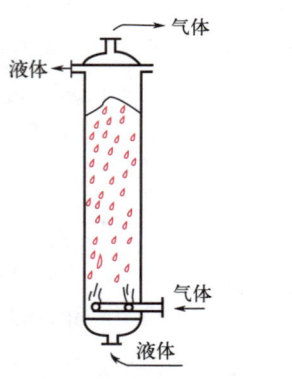

图 16-3 空心式鼓泡塔

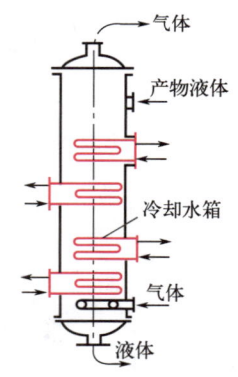

图 16-4 具有塔内热交换单元的鼓泡塔

为克服鼓泡塔中的液相返混现象，当高径比较大时，亦常采用多段鼓泡塔，以提高反应效果，见图 16-5。对于高黏性物系，例如生化工程的发酵、环境工程中活性污泥的处理、有机化工中催化加氢（含固体催化剂）等情况，常采用气体提升式鼓泡反应器（如图 16-6 所示）或液体喷射式鼓泡反应器（如图 16-7 所示），此种类型利用气体提升和液体喷射形成有规则的循环流动，可以强化反应器传质效果，并有利于固体催化剂的悬浮。

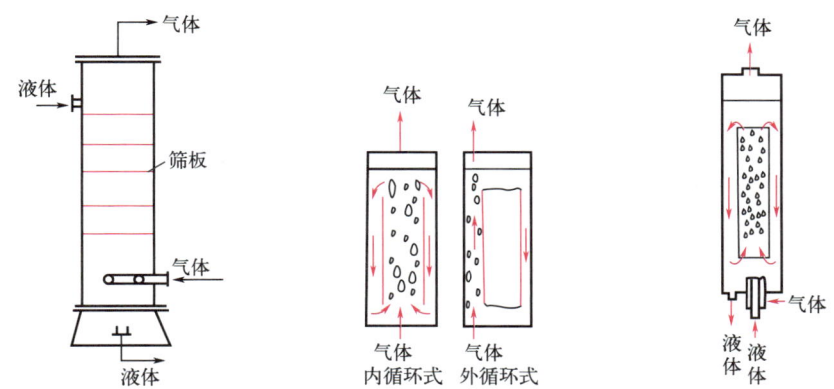

图 16-5　多段式鼓泡反应器　　图 16-6　气体提升鼓泡反应器　　图 16-7　液体喷射式鼓泡反应器

气体分布器，gas distributor，分布器的结构要求使气体均匀地分布在液层中；分布器鼓气管端的直径大小，要使鼓出来的气体泡小，使液相层中含气率增加，液层内搅动激烈，有利于气液相传质过程。图 16-8 为常见气体分布器结构。

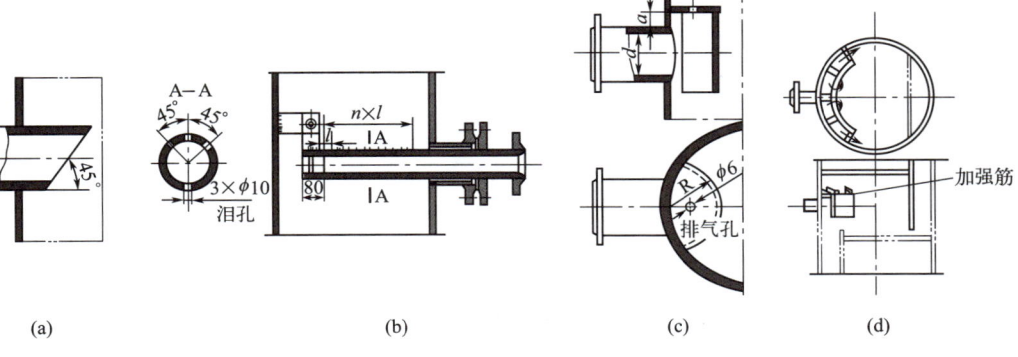

图 16-8　常见气体分布器结构

塔筒体，tower body，主要是气液鼓泡层，是反应物进行化学反应和物质传递的气液层。如果需要加热或冷却时，可在筒体外部加上夹套，或在气液层中加上蛇管。

气液分离器，gas-liquid separator，塔顶的扩大部分，内装液滴捕集装置，以分离从塔顶出来气体中夹带的液滴，达到净化气体和回收反应液的作用。常见的气液分离器结构如图 16-9 所示。

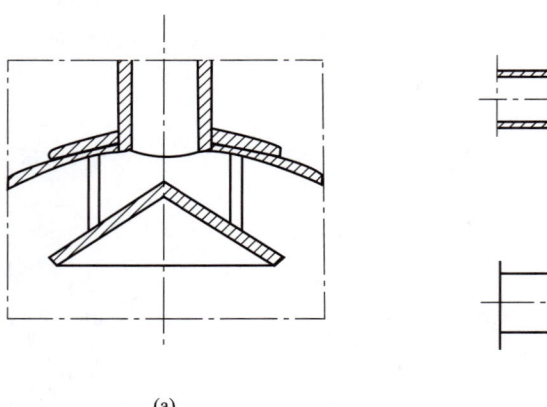

(a)

(b)

图 16-9 常见气液分离器结构

互动练习

16-6　What is a bubble column reactor?

A）A reactor used for gas-solid reactions

B）A reactor used for gas-liquid reactions

C）A reactor used for solid-liquid reactions

D）A reactor used for solid-solid reactions

16-7　What are the types of bubble column reactors?

A）Hollow type，multistage type，gas-lift type，and liquid jet type

B）Hollow type，multistage type，gas-solid type，and liquid jet type

C）Hollow type，multistage type，gas-liquid type，and liquid-solid type

D）None of the above

16-8　What is the function of the gas distributor in a bubble column reactor?

A）To distribute the liquid evenly throughout the reactor

B）To distribute the gas evenly throughout the reactor

C）To separate the gas from the liquid phase

D）None of the above

16-9　What is the column body of a bubble column reactor made of?

A）Glass only

B）Metal only

C）Plastic only

D）Different materials such as glass，metal，or plastic

16-10　What is the purpose of the gas-liquid separator in a bubble column reactor?

A）To distribute the gas evenly throughout the reactor

B）To distribute the liquid evenly throughout the reactor

C）To separate the gas from the liquid phase

D）None of the above

16.3 Structure of Packed Tower Reactors 填料塔反应器结构

Packed bed reactors（填料床反应器），also known as packed tower reactors（填料塔反应器），are widely used in the chemical industry to carry out gas-liquid and liquid-liquid reactions. The structure of a packed bed reactor includes the tower body，tower support（塔支架），manhole，handhole，and internal components. The tower body is usually cylindrical or rectangular in shape and can be made of different materials such as glass，metal，or plastic. The tower support is designed to support the weight of the tower and its contents. The manhole and handhole provide access to the inside of the tower for maintenance and inspection purposes. The internal components of a packed bed reactor include the packing material and the distribution system for the reactants.

The performance of the packing material（填料材料）in a packed bed reactor is critical to achieving efficient mass transfer and reaction. The packing material is designed to provide a large surface area for the reactants to interact and exchange mass. The choice of packing material depends on the specific reaction and the operating conditions. Common packing materials include metal，plastic，and ceramic. The performance of the packing material can be evaluated based on parameters such as the specific surface area（比表面积），porosity（空隙率），and pressure drop. The specific surface area is a measure of the amount of surface area per unit volume of packing material，while the void fraction is a measure of the empty space within the packing material. The pressure drop is the difference in pressure between the inlet and outlet of the reactor and is influenced by the size，shape，and packing density of the packing material.

In summary，the structure of a packed bed reactor includes the tower body，tower support，manhole，handhole，and internal components. The performance of the packing material is critical to achieving efficient mass transfer and reaction and can be evaluated based on parameters such as specific surface area，void fraction，and pressure drop.

技术理论

填料塔的结构较简单，如图 16-10 所示，主要包括塔体、塔体支座、人孔、手孔和塔内件。填料塔的塔身是一直立式圆筒，底部装有填料支承板，填料以乱堆或整砌的方式放置在支承板上。在填料的上方安装填料压板，以限制填料随上升气流的运动。

关键词详解

塔体，tower body，塔体是塔设备的主要部件，大多数塔体是等直径、等壁厚的圆筒体，顶盖以椭圆形封头为多。但随着装置的大型化，不等直径、不等壁厚的塔体逐渐增多。塔体除满足工艺条件对它提出的强度和刚度要求外，还应考虑风力、地震、偏心载荷所带来的影响，以及吊装、运输、检验、开停工等情况。

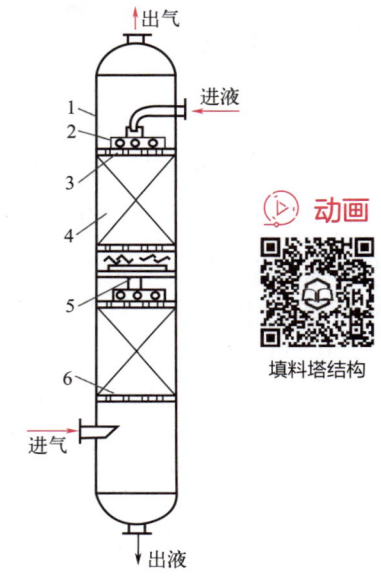

图 16-10　填料塔结构示意图

1—塔体；2—液体分布器；3—填料压紧装置；

4—填料层；5—液体收集与再分布装置；

6—支承栅板

塔体材质常采用的有非金属材料（如塑料，陶瓷等）、碳钢（复层、衬里）、不锈耐酸钢等。

塔体支座，tower support，塔设备常采用裙式支座（图 16-11）。它应当具有足够的强度和刚度，来承受塔体操作重量、风力、地震等引起的载荷。塔体支座的材质常采用碳素钢，也有采用铸铁的。

人孔，manhole，人孔是安装或检修人员进出塔器的唯一通道。人孔的设置应便于人员进入任何一层塔板。

手孔，handhole，手孔是指手和手提灯能伸入的设备孔口，用于不便进入或不必进入设备即能清理、检查或修理的场合。

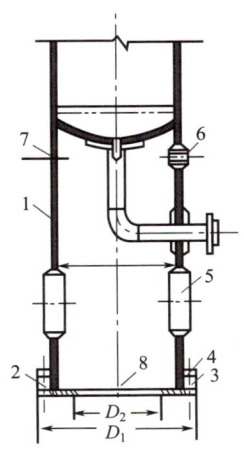

图 16-11　裙式支座

1—裙座圈；2—支承板；3—角牵板；
4—压板；5—人孔；6—有保温时排气管；
7—无保温时排气管；8—排液孔

塔内件，tower internals，填料塔的内件有填料、填料支承装置、填料压紧装置、液体分布装置、液体收集再分布装置等。合理地选择和设计塔内件，对保证填料塔的正常操作及优良的传质性能十分重要。

填料，filler，是填料塔的核心构件，它提供了气液两相接触传质的界面，是决定填料塔性能的主要因素。

填料性能的优劣通常根据效率、通量及压降三要素衡量。在相同的操作条件下，填料的比表面积越大，气液分布越均匀，表面的润湿性能越优良，则传质效率越高；填料的空隙率越大，结构越开敞，则通量越大，压降亦越低。

互动练习

16-11　What is a packed bed reactor?

A）A reactor used for gas-solid reactions

B）A reactor used for gas-liquid and liquid-liquid reactions

C）A reactor used for solid-liquid reactions

D）A reactor used for solid-solid reactions

16-12　What are the components of a packed bed reactor?

A）Tower body, tower support, manhole, handhole, and internal components

B）Tower body, tower support, gas distributor, liquid distributor, and internal components

C）Tower body, tower support, gas-liquid separator, liquid-liquid separator, and internal components

D）None of the above

16-13　What is the purpose of the packing material in a packed bed reactor?

A）To provide a large surface area for the reactants to interact and exchange mass

B）To distribute the gas evenly throughout the reactor

C）To separate the gas from the liquid phase

D）None of the above

16-14　What are the common packing materials used in a packed bed reactor?

A）Metal，plastic，and ceramic

B）Glass，metal，and plastic

C）Rubber，wood，and paper

D）None of the above

16-15　How can the performance of the packing material in a packed bed reactor be evaluated?

A）Based on the specific surface area，void fraction，and pressure drop

B）Based on the color，texture，and shape of the packing material

C）Based on the temperature，pressure，and flow rate of the reactants

D）None of the above

16.4　Selection of Gas-Liquid Phase Reactors 气液相反应器选择

Gas-liquid phase reactors are widely used in the chemical industry due to their high production capacity，improved reaction selectivity，reduced energy consumption，and better temperature control. These reactors are particularly suitable for exothermic reactions（放热反应）that require efficient heat removal to maintain the desired reaction temperature. They are also useful for carrying out reactions that involve gas-liquid mass transfer，such as gas absorption，stripping（气液分离），and catalytic reactions.

In addition to the above advantages，gas-liquid reactors can be operated with low liquid flow rates，resulting in significant savings in operating costs. This is because the reactant gases can be effectively dispersed in the liquid phase，providing a large interfacial area for the reaction to occur. The high interfacial area also allows for efficient mass transfer between the gas and liquid phases，which promotes reaction rates and product yields.

When selecting a gas-liquid reactor，it is important to consider factors such as reactor type，material of construction，operating conditions，and the nature of the reaction. The choice of reactor type will depend on the specific reaction requirements，such as the need for a high surface area-volume ratio （面积体积比），efficient gas-liquid mixing，or the ability to handle corrosive or high-pressure conditions. Proper reactor selection can help to optimize the reaction conditions and maximize the desired product yields.

技术理论

可用于气液相反应过程的反应器类型较多，选择时一般应考虑以下因素：具备较高的生产能力，有利于提高反应选择性、降低能量消耗、控制反应温度，能在较少液体流率下操作。

关键词详解

放热反应，exothermic reactions，在化学反应中，反应物总能量大于生成物总能量的反应叫做放热反应。包括燃烧、中和、金属氧化、铝热反应、较活泼的金属与酸反应、由不稳定物质变为稳定物质的反应。

互动练习

16-16　What are the advantages of using gas-liquid reactors in chemical industry?

A）High production capacity

B）Improved reaction selectivity

C）Reduced energy consumption

D）All of the above

16-17　In which type of reactions are gas-liquid reactors particularly suitable?

A）Endothermic reactions

B）Reactions with low interfacial area

C）Exothermic reactions requiring efficient heat removal

D）None of the above

16-18　What is the significance of low liquid flow rates in gas-liquid reactors?

A）It results in significant savings in operating costs

B）It promotes efficient gas-liquid mixing

C）It is useful for endothermic reactions

D）None of the above

16-19　What factors should be considered when selecting a gas-liquid reactor?

A）Reactor type

B）Material of construction

C）Operating conditions

D）All of the above

16-20　How does proper reactor selection help to optimize the reaction conditions?

A）It maximizes the desired product yields

B）It minimizes the energy consumption

C）It eliminates the need for temperature control

D）None of the above

Design of Bubble Column Reactors
鼓泡塔反应器设计

　　本任务重点讲解鼓泡塔反应器的设计原理及关键参数（如气液比、塔径等）。读者将学习如何通过优化设备结构和运行条件提升传质效率，并降低生产成本。

学习目标

知识目标

　　（1）理解鼓泡塔反应器的基本工作原理及特性。

　　（2）掌握鼓泡塔反应器设计中的关键参数，如气液比、塔径和高度。

　　（3）了解鼓泡塔反应器在工业中的典型应用场景。

技能目标

　　（1）能根据工艺需求设计鼓泡塔反应器的结构和尺寸。

　　（2）能优化鼓泡塔反应器的运行条件以提高传质效率。

价值目标

　　（1）注重鼓泡塔反应器设计中的环保和节能。

　　（2）强化设备设计与工业需求的适配性。

17.1　Fundamentals of Gas-Liquid Phase Reaction Kinetics
气液相反应动力学基础

　　Gas-liquid phase reactions are characterized by the transfer of reactants from the gas phase into the liquid phase，where they undergo chemical transformations（化学转化）. The rate of these reactions can be expressed in terms of the volumetric mass transfer coefficient（体积传质系数）or the interfacial area（界面面积）. The volumetric mass transfer coefficient is a measure of the rate of mass transfer of gas into the liquid，whereas the interfacial area represents the surface area per unit volume of the gas-liquid interface.

The macroscopic kinetics（宏观动力学）of gas-liquid reactions are influenced by several factors，including the gas-liquid interfacial area，the concentration of reactants in the gas and liquid phases，the mass transfer rate，and the reaction rate. The basic characteristics of gas-liquid reactions include a high interfacial area，fast mass transfer，and strong coupling between mass transfer and reaction kinetics.

The theory of gas-liquid mass transfer provides a theoretical framework（理论框架）for understanding gas-liquid reactions. This theory is based on the concept of interfacial area and assumes that the rate of mass transfer is proportional to the concentration gradient across the interface. The macroscopic kinetics of gas-liquid reactions can be described by a set of differential equations（微分方程）that relate the concentration of reactants and products to time，and take into account the mass transfer and reaction rates.

In summary，the understanding of the macroscopic kinetics of gas-liquid reactions is crucial for the design and optimization of gas-liquid reactors. The application of mass transfer and reaction kinetics theories can aid in the development of more efficient and effective gas-liquid reactors.

技术理论

化学反应速率是指在反应系统中，某一物质在单位时间、单位反应区域内的反应量。如式(2-1)所示。均相反应过程的反应区域通常取反应混合物总体积，则反应速率单位以 $kmol/(m^3 \cdot h)$ 表示。

$$反应速率 = \frac{反应量}{反应区域 \times 反应时间} \tag{2-1}$$

式中的反应区域，对于气液相反应过程有以下几种选择：

① 选用液相体积时，反应速率（$-r_A$）单位为 $kmol/(m^3\ 液体 \cdot h)$；

② 选用气液相混合物体积时，反应速率（$-r_A$）$_V$ 单位为 $kmol/(m^3\ 气液相混合物 \cdot h)$；

③ 选用单位气液相界面积时，反应速率（$-r_A$）$_S$ 单位为 $kmol/(m^2\ 相界面 \cdot h)$。

气液相反应系统中，单位液相体积所具有的气液相界面积为

$$a_i = \frac{相界面积}{液相体积} = \frac{S}{V_L}$$

而单位气液混合物体积所具有的气液相界面积

$$a = \frac{相界面积}{气液相混合物体积} = \frac{S}{V_R} = \frac{S}{V_G + V_L}$$

a_i 和 a 均称为比相界面，但它们的基准不同，故数值上也有差别。两者可以通过气含率 ε 关联。

气含率的定义为：单位气液混合物体积中气相所占的体积分数，即

$$\varepsilon = V_G/V_R = V_G/(V_G + V_L) \tag{17-1}$$

根据其定义，可得到如下关系

$$a = (1-\varepsilon)a_i \tag{17-2}$$

气液相反应过程的三种反应速率有如下关系

$$(-r_A)_V = (-r_A)(1-\varepsilon) = (-r_A)_S a \tag{17-3}$$

气液相反应的基本特征可归纳成以下三点。

① 无论在液相中进行的是简单反应还是复杂反应，宏观上总可以将气液相反应分解成传质和反应两个过程，这两个过程组成一个统一体，先传质后反应。

② 传质和反应的统一体内，传质和反应双方互相影响和制约。这个统一体所表现出来的速率，往往既非反应的本征速率，也非传质的本征速率，而是这两者矛盾统一的速率——宏观速率。

③ 传质和反应统一体的统一水平受流体力学、传热和传质等传递过程和流体的流动与混合等因素的影响。这个统一水平是相对的、可以变化的，即可调的。

气液相界面物质传递的模型有多个，如"双膜理论""表面更新理论""渗透理论"等。但应用最广的是路易斯-卫特曼（Lewis-Whitman）于1923年提出的"双膜理论"，其优点是简明易懂，便于进行数学处理。

设有二级不可逆气液相反应

$$A(气相) + bB(液相) \longrightarrow C(产物)$$

气相组分 A 与液相组分 B 之间的反应过程，需要经历以下步骤。

① 气相组分 A 从气相主体传递到气液相界面，在界面上假定达到气液相平衡。

② 气相组分 A 从气液相界面扩散入液相，并在液相内进行化学反应。

③ 液相内的反应产物向浓度梯度下降的方向扩散，气相产物则向界面扩散。

④ 气相产物向气相主体扩散。

气液相反应，根据不同的传质速率和化学反应速率，有八种不同的反应类型：瞬间反应、界面反应、二级快速反应、拟一级快速反应、二级中速反应、拟一级中速反应、二级慢速反应和极慢速反应。

关键词详解

简单反应，simple reaction，反应速率可表示为诸反应物浓度的正整数方幂的单一反应。

双膜理论，bimembrane theory，假设平静的气液界面两侧存在着气膜与液膜，是很薄的静止层或层流层，如图 17-1 所示。当气相组分向液相扩散时，必须先到达气液相界面，并在相界面上达到气液平衡，即服从亨利定律

$$p_{A_i} = H_A c_{A_i} \qquad (17-4)$$

式中，p_{A_i} 为气相组分 A 在相界面上成平衡的气相分压，Pa；c_{A_i} 为气相组分 A 在相界面上成平衡的液相浓度，kmol/m³；H_A 为亨利常数，m³·Pa/kmol。

双膜模型又假设在气膜之外的气相主体和液膜之外的液相主体中达到完全的混合均匀，即全部传质阻力都集中在膜内。

瞬间反应，instantaneous reaction，气相组分 A 与液相组分 B 之间的反应为瞬间完成，两者不能共存，反应发生于液膜内某一个面上，该面称为反应面。在反应面上 A、B 的浓度均为零。所以，A 和 B 扩散到此界面的速率决定了过程的总速率。

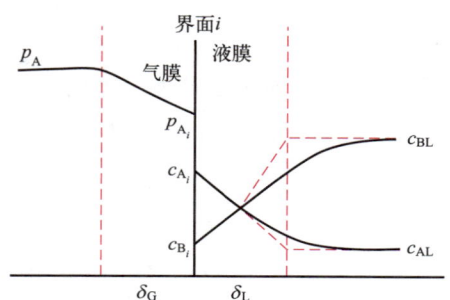

图 17-1 气液相反应双膜理论模型

界面反应，interface reaction，反应的性质与瞬间反应相同，但因液相中组分 B 的浓度 c_B 高，气相组分 A 一扩散到达界面即反应完毕，反应面移至相界面上。在界面上，A 组分浓度为零，而 B 组分浓度可大于零。此时，总反应速率取决于气膜内 A 的扩散速率。

二级快速反应，secondary rapid reaction，相当于瞬间反应的反应面扩展成为一个反应区，在反应区内 A、B 并存。但由于尚属于快反应，反应区仍在液膜内，并不进入液相主体。

拟一级快速反应，quasi first level rapid reaction，与二级快速反应一样，反应发生于液膜内某一区域中。但组分 B 的浓度 c_B 高，以致与 A 发生反应后消耗的量可以忽略不计，故可视为拟一级反应，即液膜内 c_B 的变化可以忽略。

二级中速反应，secondary medium speed reaction，A 与 B 在液膜中发生反应，但因反应速率不是很快，故有部分 A 在液膜中不能反应完毕，因而进入液相主体，并在液相主体中继续与 B 组分反应。

拟一级中速反应，pseudo first order medium speed reaction，与二级中速反应一样，反应同时发生于液膜与液相主体中。但因液相中 B 组分浓度高，使得在整个液膜中 B 的浓度近似不变，成为 A 组分的拟一级反应。

二级慢速反应，secondary slow reaction，与传质速率相比，A 与 B 的反应很慢，扩散通过相界面的气相组分 A 在液膜中与液相组分 B 发生反应，但大部分 A 反应不完而扩散进入液相主体，并在液相主体中与 B 发生反应。由于液膜在整个液相中所占体积分数很小，故反应主要在液相主体中进行。

极慢速反应，extremely slow reaction，A 与 B 的反应极其缓慢，传质阻力可以忽略，在液相中组分 A 和 B 是均匀的，反应速率完全取决于化学反应动力学。

互动练习

17-1　What is the measure of the rate of mass transfer of gas into the liquid in gas-liquid reactions?

A）The interfacial area

B）The reaction rate

C）The concentration gradient

D）The volumetric mass transfer coefficient

17-2　What are the basic characteristics of gas-liquid reactions?

A）Low interfacial area，slow mass transfer，and weak coupling between mass transfer and reaction kinetics

B）High interfacial area，fast mass transfer，and strong coupling between mass transfer and reaction kinetics

C）Low interfacial area，fast mass transfer，and strong coupling between mass transfer and reaction kinetics

D）High interfacial area，slow mass transfer，and weak coupling between mass transfer and reaction kinetics

17-3　What is the theory of gas-liquid mass transfer based on?

A）The reaction rate

B）The concentration gradient

C）The interfacial area

D）The mass transfer coefficient

17-4　What do the differential equations used to describe the macroscopic kinetics of gas-liquid reactions take into account?

A）Only the reaction rate

B）Only the mass transfer rate

C）Both the mass transfer and reaction rates

D）Only the concentration gradient

17-5　Why is the understanding of the macroscopic kinetics of gas-liquid reactions crucial for the design and optimization of gas-liquid reactors?

A）It allows for a better understanding of the interfacial area

B）It helps to determine the type of reactor to use

C）It enables the development of more efficient and effective gas-liquid reactors

D）It is not important for the design and optimization of gas-liquid reactors

17.2 Transport Characteristics of Bubble Column Reactors 鼓泡塔反应器传递特性

Bubble column reactors is a widely used type of gas-liquid reactor in chemical industries. The hydrodynamics（流体动力学）of bubble columns play a crucial role in reactor performance. The liquid flow regime（液体流动状态）and bubble characteristics such as size，shape，and distribution can affect mass and heat transfer efficiency.

Bubble column reactors have low pressure drops due to their uniform flow patterns and are resistant to clogging（防止堵塞）. The gas-liquid interface area（气液相界面面积），gas hold-up（气含率），and mixing time are essential parameters to determine the mass transfer rate. The interfacial area is the area of contact between the two phases and the gas hold-up is the fraction of the volume of gas in the column. The liquid side mass transfer coefficient（液相侧传质系数）can be increased by adjusting the bubble size and gas flow rate.

Heat transfer in bubble columns is primarily affected by the liquid phase，and the gas phase plays a minor role. Bubble columns can achieve high heat transfer rates by using high-velocity liquids. The design of the bubble column is crucial to minimize backmixing and maximize mass and heat transfer.

技术理论

鼓泡塔传递特性包括流动状态和气泡特性、气泡大小、气含率、气液比相界面积、鼓泡塔内的气体阻力和返混。

关键词详解

流动状态和气泡特性，flow state and bubble characteristics，工业鼓泡塔反应器通常在两种流动状态下操作，即安静区和湍动区。所谓安静区操作，即鼓泡塔中的气体流量较小，气泡大小比较均匀，气泡规则地浮升，液体搅拌并不显著。在安静区操作，既能达到一定的气体流量，又可避免气体的轴向返混，很适用于动力学控制的慢反应。对于典型的气液体系（空气-水体系），安静区的空塔气速 u_{OG} 通常小于 0.05m/s，气体分布器的孔口气速小于 7m/s，此时在气体分布器孔口直接形成气泡，其气泡的形状、大小和运动与孔口的直径有关。孔径很小时（如 1mm），形成球形气泡螺旋上升，气泡直径小于 2mm；孔径较大时（如 2mm），形成当量直径约为 3～6mm 的椭圆形气泡，上升过程中左右摆动；孔径大时（如 4mm），形成当量直径大于 6mm 的菌帽形气泡，具有明显的尾涡。显然，在安静区操作的鼓泡塔，其气体分布器的设计十分重要。一般常采用多孔板或多孔盘管，孔径小于 3mm，开孔率一般也小于 5%。

在气体流量较大时，气泡运动呈不规则现象，液体高度地湍动，塔内物料强烈混合，气泡作用的机理比较复杂，这种情况称为湍动区。在湍动区气泡大小不均匀，大气泡上升速度快，小气泡上升速度慢，停留时间不等，加之无定向搅动，不仅呈极大的液相返混，也造成气相返混。湍动区的空塔气速 u_{OG} 通常大于 0.08m/s，工业上常采用大孔径的单管或特殊型式的喷嘴作为气体分布装置。气泡不是在分布器孔口处形成，而是在孔口处形成一股气流，气泡是靠气流与液体之间的喷射、冲击和摩擦而形成。因此在这种鼓泡塔内气泡的形状、大小和运动是各式各样的，是瞬息万变的，是随机的，形成大小不一的气泡群。

气泡大小，bubble size，气泡的大小直接关系到气液传质面积。在同样的空塔气速下，气泡越小，说明分散越好，气液相接触面积就越大。在安静区，因为气泡上升速度慢，所以小孔气速对其大小影响不大，主要与分布器孔径及气液特性有关。

气含率，gas hold-up，是气液混合液中气体所占的体积分数，可用式（17-5）表示

$$\varepsilon_G = \frac{V_G}{V_L + V_G} = \frac{V_G}{V_{GL}} \qquad (17\text{-}5)$$

式中，ε_G 为气含率；V_G 为气体体积，m^3；V_L 为液体体积，m^3；V_{GL} 为气液混合物体积，m^3。

影响气含率的因素主要有设备结构、物性参数和操作条件等。

气液比相界面积，gas-liquid specific interfacial area，是指单位气液混合鼓泡床层体积所具有的气泡表面积，可以通过气泡平均直径 d_{VS} 和气含率 ε_G 计算出，即

$$a = \frac{6\varepsilon_G}{d_{VS}} \quad (m^2/m^3) \qquad (17\text{-}6)$$

鼓泡塔内的气体阻力，gas resistance inside the bubbling column，鼓泡塔内的气体阻力 Δp 由两部分组成：一是气体分布器阻力，二是床层静压头的阻力。即

$$\Delta p = \frac{10^{-3}}{C^2} \frac{u_0^2 \rho_G}{2} + H_{GL}\rho_{GL}g \quad (Pa) \qquad (17\text{-}7)$$

式中，C^2 为小孔阻力系数，约为 0.8；u_0 为小孔气速，m/s；ρ_{GL} 为鼓泡层密度，kg/m^3。

返混，back mixing，鼓泡塔内液相存在返混，所以通常工业鼓泡塔反应器内液相视为理想混合。塔内气体的返混一般不太明显，常假设为置换流，其计算误差约为 5%。但要求严格计算时，尤其是当气体的转化率较高时，需考虑返混。

互动练习

17-6　What is the main factor that affects mass and heat transfer efficiency in bubble column reactors?

A）Reactor diameter

B）Liquid flow regime and bubble characteristics

C）Gas flow rate

D）Residence time

17-7　What is the advantage of bubble column reactors in terms of pressure drops?

A）They have high pressure drops

B）They have low pressure drops

C）They are prone to clogging

D）They have uneven flow patterns

17-8　　What are the essential parameters to determine mass transfer rate in bubble column reactors?

A）Reactor diameter，gas flow rate，and residence time

B）Liquid side mass transfer coefficient，bubble size，and gas flow rate

C）Reactor height，gas hold-up，and mixing time

D）Liquid phase velocity，interfacial area，and gas hold-up

17-9　　What is the primary factor affecting heat transfer in bubble columns?

A）Gas flow rate

B）Residence time

C）Liquid phase velocity

D）Liquid phase

17-10　　What is the main objective of bubble column reactor design?

A）To minimize clogging

B）To maximize backmixing

C）To maximize mass and heat transfer

D）To minimize gas hold-up

17.3　Design of Bubble Column Reactors 鼓泡塔反应器设计

Bubble column reactors are widely used in chemical processes and are characterized by the presence of gas bubbles that rise through a liquid phase. The design of these reactors requires careful consideration of several factors，including the determination of reactor diameter，height，and volume.

Empirical methods are often used to estimate the reactor diameter，which is typically between 0. 3～2 meters. The reactor height is usually chosen based on the required residence time，with a typical value of 2～6 meters. The reactor volume is then calculated using the chosen diameter and height.

However，empirical methods have limitations，and a more accurate approach involves the use of mathematical models. The mathematical model for bubble column reactors is based on the conservation of mass and energy equations（质量守恒和能量守恒方程）. The model considers the interactions between the gas and liquid phases，as well as the reaction kinetics.

The design of a bubble column reactor using a mathematical model involves determining the appropriate values for the gas hold-up，liquid phase velocity，and mass transfer coefficient. These values are then used to calculate the reactor performance，such as the conversion rate and the yield of the desired product.

Overall，the design of a bubble column reactor involves a careful consideration of various factors，and the use of empirical methods and mathematical models can provide useful insights into reactor performance.

技术理论

鼓泡塔反应器工艺设计计算的主要内容是气液鼓泡床的体积计算。对半连续操作的鼓泡塔反应器体积的计算，可归结为反应时间的计算，这与均相间歇操作的反应器计算类似。对连续操作的鼓泡塔反应器体积的计算，往往归结为鼓泡层高度的确定。

鼓泡塔反应器工艺设计计算的方法主要有经验法和数学模型法。

当缺乏宏观动力学数据时，无法进行数学模型法计算，此时可用比较简便的经验法解决。经验法主要包括反应器直径的确定、反应器高度和体积的经验确定。

关键词详解

反应器直径的确定，determination of reactor diameter，鼓泡塔反应器的直径由最佳空塔气速来决定。决定了最佳空塔气速也就确定了鼓泡塔反应器的最佳高度和直径。

最佳空塔气速应满足两个条件：①保证反应过程的最佳选择性；②保证反应器体积最小。

u_{OG} 的实际最佳值确定之后，鼓泡塔反应器直径 D 可按式(17-8) 计算

$$D = \sqrt{\frac{4v_G}{\pi u_{OG}}} \qquad (17-8)$$

式中，v_G 为气体体积流量，m^3/h。

反应器高度和体积的确定，determination of reactor height and volume，除充气床层体积（即反应器的有效体积）外，鼓泡塔体积还包括充气液层上部除沫分离空间体积和反应器顶盖的死区体积。反应器高度的确定，应全面考虑床层含气量、雾沫夹带、床层上部气相的允许空间（有时为了防止气相爆炸，要求空间尽量小一些）、床层出口位置和床层液面波动范围等多种因素的影响而后确定。

互动练习

17-11　What is the typical range for the reactor diameter in bubble column reactors?

A）0.1～0.3 meters

B）0.3～2 meters

C）2～6 meters

D）6～10 meters

17-12　What is the main limitation of empirical methods in designing bubble column reactors?

A）They are too complex to use

B）They do not consider reaction kinetics

C）They do not provide accurate values for reactor height

D）They are not applicable to mathematical models

17-13　What is the main advantage of using mathematical models in bubble column reactor design?

A）They are easier to use than empirical methods

B）They provide accurate values for reactor height

C）They consider reaction kinetics

D）They are not applicable to gas hold-up

17-14　What is the main factor used to determine reactor height in bubble column reactors?

A）Gas hold-up

B）Liquid phase velocity

C）Required residence time

D）Mass transfer coefficient

17-15　What is the purpose of calculating reactor performance in bubble column reactor design?

A）To determine the appropriate values for gas hold-up and mass transfer coefficient

B）To estimate the reactor diameter

C）To determine the appropriate values for liquid phase velocity

D）To evaluate the conversion rate and yield of the desired product

任务18

Design of Packed Tower Reactors
填料塔反应器设计

任务要点

　　本任务讲解填料塔反应器的设计方法及填料选择原则，涵盖压降、气液分布等关键设计参数。读者将学习如何设计高效的填料塔反应器，并解决设备运行中的常见问题。

学习目标

知识目标

　　（1）了解填料塔反应器的基本结构及原理。

　　（2）掌握填料塔反应器设计中的关键要素，如填料类型选择和压降计算。

　　（3）认识填料塔反应器在化工生产中的典型应用。

技能目标

　　（1）能选择合适的填料并设计填料塔反应器的结构和操作条件。

　　（2）能解决填料塔反应器运行中常见的问题，如堵塞和压降过大。

价值目标

　　（1）提高填料塔反应器的运行效率和经济性。

　　（2）培养对填料塔反应器工艺优化和环保设计的关注。

　　Packed tower reactors（填料塔反应器）are widely used in the chemical industry due to their high mass transfer efficiency and flexible design. The design of a packed tower reactor involves determining the column diameter and bed height to achieve the desired conversion or reaction rate. The diameter of the column is typically determined by the superficial gas velocity（表观气体速度），while the bed height is determined by the required residence time（必要停留时间）.

　　The packing material used in the reactor plays a crucial role in determining the mass transfer and reaction rate. The mass transfer coefficient can be improved by selecting a suitable packing material and optimizing the flow conditions. In addition to mass transfer efficiency，the pressure drop across the packed bed is an important consideration in the design. A higher bed height results in a higher pressure drop，which can affect the overall reactor performance.

The use of multiple beds in sequential or parallel can help to reduce the pressure drop while maintaining high mass transfer efficiency. The design of packed tower reactors can be optimized through experiments or numerical simulations（实验或数值模拟），taking into account the reaction kinetics，thermodynamics，and fluid dynamics of the system.

技术理论

填料塔反应器的工艺设计计算主要包括填料塔反应器的塔径与填料层高度的计算。

关键词详解

反应器塔径的计算，calculation of reactor tower diameter，由埃克特图（或方程）计算出液泛气速，再取液泛气速的 0.6～0.8 倍作为填料塔式反应器的空塔气速，由此可计算出塔径，公式为：

$$D = \sqrt{\frac{4v_G}{\pi u_{OG}}} \tag{17-8}$$

填料层高度的计算，calculation of the height of the filling layer，塔高（填料层高度）的计算公式可从塔内微元高度 $\mathrm{d}Z$ 作物料衡算求得，设气体内被吸收溶质的含量较少

$$G\mathrm{d}y_A = K_{GA}a(y_A - y_A^*)p\mathrm{d}Z \tag{18-1}$$

式中，G 为气体的空塔摩尔流速，$\mathrm{mol}/(\mathrm{m}^2 \cdot \mathrm{s})$；$y_A$、$y_A^*$ 分别为气相中被吸收组分 A 的摩尔分数和液流主体中 A 的平衡摩尔分数；a 为单位体积气-液传质比表面积，$\mathrm{m}^2/\mathrm{m}^3$；$K_{GA}$ 为以气相量表示的总传质系数，$\mathrm{mol}/(\mathrm{m}^2 \cdot \mathrm{Pa} \cdot \mathrm{s})$。它与气膜传质系数 k_{GA}、液膜传质系数 k_{LA} 的关系为

$$\frac{1}{K_{GA}a} = \frac{1}{k_{GA}a} + \frac{H_A}{k_{LA}a} \tag{18-2}$$

对式（18-2）分离变量积分，得

$$Z = \frac{G}{p}\int_{y_{A2}}^{y_{A1}} \frac{\mathrm{d}y_A}{K_{GA}a(y_A - y_A^*)} \tag{18-3}$$

式中，y_{A1}、y_{A2} 分别为进、出吸收塔气体中被吸收组分 A 的摩尔分数。

式(18-3) 一般用图解积分法求解。如果平衡线为直线（$y^* = mx$），平均推动力取 $y_A - y_A^*$ 的对数平均值，式(18-3) 可简化成

$$Z = \frac{G}{(K_{GA}a)_m p} \times \frac{y_{A1} - y_{A2}}{(y_A - y_A^*)_{LM}} \tag{18-4}$$

式中，下标 m 和 LM 分别表示平均值和对数平均值。$(K_{GA}a)_m$ 一般指塔底和塔顶的算术平均值。

如果气体内被吸收溶质的含量较高，且大部分被吸收，式(18-1) 内的 G 不再为定值，应该用惰性气体量为基准进行物料衡算。

$$G_I d\left(\frac{y_A}{1 - y_A}\right) = K_{GA}a(y_A - y_A^*)p\,dZ \tag{18-5}$$

或

$$Z = \frac{G_I}{p}\int_{y_{A2}}^{y_{A1}} \frac{dy_A}{K_{GA}a(1 - y_A)^2(y_A - y_A^*)} \tag{18-6}$$

式中，G_I 为惰性气体的恒摩尔流速，$mol/(m^2 \cdot s)$。其他参数的意义同式(18-1)。

上式亦可用图解积分法求 Z，对 $1 - y_A$ 和 $K_{GA}a$ 值用进、出口的平均值，上式可简化成：

$$Z = \frac{G_I}{p(K_{GA}a)_m(1 - y_A)_m^2}\int_{y_{A2}}^{y_{A1}} \frac{dy_A}{y_A - y_A^*} \tag{18-7}$$

如果平衡线为直线，可得到与式(18-4) 相一致的计算式

$$Z = \frac{G}{(K_{GA}a)_m p(1 - y_A)_m^2} \times \frac{y_{A1} - y_{A2}}{(y_A - y_A^*)_{LM}} \tag{18-8}$$

通常 y_A^* 可以忽略不计，从式(18-6) 可得

$$Z = \frac{G_I}{(K_{GA}a)_m p}\left\{\frac{1}{1 - y_{A1}} - \frac{1}{1 - y_{A2}} - \ln\left[\left(\frac{1 - y_{A1}}{y_{A1}}\right)\left(\frac{1 - y_{A2}}{y_{A2}}\right)\right]\right\} \tag{18-9}$$

以上公式是取 $K_{GA}a$ 在塔内的平均值。但是，在塔内气体流量有很大变化时，$K_{GA}a$ 值会有显著的变化，因而用以上公式计算有较大的误差。

在式(18-2) 中 k_{GA} 与 k_{LA} 分别是 A 组分的气膜及液膜传质系数 [单位分别是 $mol/(s \cdot m^2 \cdot Pa)$ 及 m/s]，根据实验测定，其值可通过以下两式计算得到：

$$\frac{k_{GA}p}{G_M} = \frac{5.23}{M}(a_t d)^{-1.7}\left(\frac{G_M d_0}{\mu_G}\right)^{-0.3}\left(\frac{\mu_G}{\rho_G D_{GA}}\right)^{-2/3} \tag{18-10}$$

$$k_{LA}\left(\frac{\rho_L}{\mu_L g}\right) = 0.005\left(\frac{G_L}{a_w \mu_L}\right)^{2/3}\left(\frac{\mu_L}{\rho_L D_{LA}}\right)^{-1/2}(a_t d)^{0.4} \tag{18-11}$$

式中，G_M、G_L 分别为气体、液体质量流速，kg/(m²·s)；M 为气体的平均摩尔量，kg/kmol；d 为填料公称直径，m；a_t 为填料层的比表面积，m²/m³；D_{GA}、D_{LA} 分别为 A 组分在气体和液体中的分子扩散系数，mol/(s·m·Pa)。

a_W 为单位填料堆积体积内的浸润面积，其与填料层的比表面积 a_t 有如下关系

$$\frac{a_W}{a_t} = 1 - \exp\left[-1.45\left(\frac{\sigma_C}{\sigma_L}\right)^{0.75}\left(\frac{G_L}{a_t\mu_L}\right)^{0.1}\left(\frac{G_L^2 a_t}{\rho_L g}\right)^{-0.05}\left(\frac{G_L^2}{\rho_L a_t\sigma_L}\right)^{0.2}\right]$$

（18-12）

式中，σ_C 为液体临界表面张力，N/m。

互动练习

18-1   What is the primary advantage of packed tower reactors?

A）High pressure drops

B）Low mass transfer efficiency

C）Flexible design

D）Low residence time

18-2   What is the determining factor for the diameter of a packed tower reactor?

A）Residence time

B）Gas velocity

C）Packing material

D）Pressure drops

18-3   How can the mass transfer coefficient in a packed bed reactor be improved?

A）By increasing the bed height

B）By selecting a suitable packing material

C）By reducing the flow rate

D）By reducing the gas velocity

18-4   What is the effect of a higher bed height on a packed tower reactor?

A）Higher mass transfer efficiency

B）Lower pressure drops

C）Lower residence time

D）Higher pressure drops

18-5　What factors should be considered when optimizing the design of a packed tower reactor?

A）Reaction kinetics and thermodynamics only

B）Fluid dynamics only

C）Experimentation only

D）Reaction kinetics，thermodynamics，and fluid dynamics

任务19

Operation and Control of Bubble Column Reactors
鼓泡塔反应器操作与控制

任务要点

本任务讲解鼓泡塔反应器的操作与控制方法，读者将学习如何优化设备运行条件，提高气液接触效率，并掌握常见问题的诊断与解决方法。

学习目标

知识目标

（1）掌握鼓泡塔反应器操作中的关键控制变量，如气液流速和温度。

（2）了解常见的操作问题及其解决方案。

技能目标

（1）能监控并调节鼓泡塔反应器的运行参数以确保产品质量。

（2）能制定应急方案以应对设备异常情况。

价值目标

（1）注重生产过程中设备安全性和稳定性的提高。

（2）培养团队协作和快速反应能力。

19.1 Operation and Control of the Alkylation Tower for Ethylbenzene Production
烃化塔操作与控制

Bubble column reactors are commonly used in the production of various chemicals，including the production of ethylbenzene from ethylene and benzene. The reaction is catalyzed by a zeolite catalyst and takes place in the presence of a solvent，such as cyclohexane. The process involves the continuous circulation of the reactants and solvent through the reactor，which is operated at high temperature and pressure.

During normal operation，the reactor is fed with a continuous flow of ethylene and benzene along with the solvent，which is circulated by a pump. The gas flow rate and reactor temperature are controlled to maintain the desired conversion and selectivity. The reaction product is continuously removed from the bottom of the reactor and sent for further processing.

To stop the reactor，the feed flow and pump are shut down，and the reactor is allowed to cool down to a safe temperature. Before restarting the reactor，the catalyst is regenerated by burning off any accumulated coke deposits.

During normal operation，the gas flow rate and reactor temperature should be monitored closely to ensure optimal performance. The gas flow rate can be adjusted by changing the gas pressure，while the reactor temperature is controlled by adjusting the heating or cooling rate.

Common operational issues in a hydrocarbon reactor（碳氢化合物反应器）include catalyst deactivation，fouling of the reactor walls or internals，and pressure build-up due to excessive coke deposition（积炭）. These issues can be mitigated by proper monitoring and maintenance of the reactor internals（反应器内部），regular cleaning and catalyst replacement，and prompt troubleshooting of any abnormalities.

In the event of a reactor emergency，such as a sudden pressure surge or loss of coolant flow，the reactor should be shut down immediately and appropriate emergency procedures followed to prevent any harm to personnel or equipment.

技术理论

乙苯生产烃化反应工艺流程如图 19-1 所示。精苯由苯贮槽用苯泵送入烃化塔，乙烯气经缓冲器送入烃化塔，根据反应的实际情况用乙烯间歇地将三氯化铝催化剂从三氯化铝槽定量地压入烃化塔。

苯和乙烯在三氯化铝槽催化剂的存在下起反应，烃化塔内的过量苯蒸气及未反应的乙烯气经过捕集器捕集，使带出的烃化液回至烃化液沉降槽，其余气体进入循环苯冷凝器中冷凝。从烃化塔出来的流体经气液分离器后，回收苯送入水洗塔，分离出来的尾气（即 HCl 气体）进入尾气洗涤塔洗涤。沉降槽上层烃化液流入烃化液缓冲罐，进入缓冲罐的烃化液由于烃化系统本身的压力压进

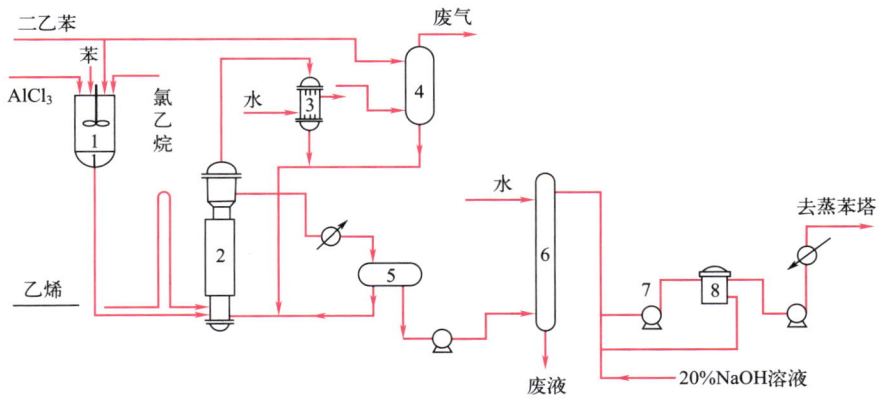

图 19-1　乙苯生产烃化反应工艺流程图

1—催化剂配制槽；2—鼓泡塔反应器；3—冷凝器；4—二乙苯吸收器；5—沉降槽；

6—水洗塔；7—中和泵；8—油碱分离槽

水洗塔底部进口，水洗塔上部出口溢出的烃化液进入烃化液中间槽，水洗塔中的污水由底部排至污水处理系统。由烃化液中间罐出来的烃化液与由碱液罐出来的 NaOH 溶液一起经过中和泵混合中和。中和之后的混合液进入油碱分离沉降槽沉降分离。

关键词详解

乙苯生产鼓泡塔反应器的开车，startup of the bubble column reactor for ethylbenzene production，首先，通入空气对系统进行试漏，保证无泄漏。组织开车人员全面检查本系统全部处于完善备用状态。与调度联系水、电、汽及其他原料。开启乙烯缓冲罐，使乙烯罐内充乙烯。开蒸汽总阀，使车间总管上有蒸汽。从催化剂计量槽压一定量催化剂进入烃化塔。用中和泵抽新碱液入第一油碱分离器，稍开乙烯阀，向塔内通乙烯，按照塔内温度上升速率控制乙烯入烃化塔流量，并注意尾气压力和尾气塔中洗涤情况。根据通入乙烯后反应情况和夹套加热，可调节蒸汽量和冷却水量。再开苯泵，稳定泵压，开泵流量计调节苯进料流量，并加大乙烯流量，根据温度情况反复调节，保证温度在 95℃ 左右，并且苯量是乙烯量的 8～10 倍。反应过程中，每小时向塔内压入新 AlCl₃ 复合体一次。经常巡回，根据设备、管道的温度估计烃化塔出料情况。

正常操作控制，normal operation and control，乙苯的反应正常操作控制过程包括温度控制、压力控制和流量控制。

烃化温度，烃化温度的高低直接影响产品的质量。温度过高时，深烃化物量增多，选择性下降；温度过低时反应速率减小，产量下降。通常维持烃化温度在 95℃±5℃ 的范围内。生产中常采用三种方法来控制反应温度：第一种方法是控制苯进量，由于该烃化反应是放热反应，当反应温度偏高时可以减小进苯量，反之则增大进苯量；第二种方法是采用向烃化塔外夹套通入水蒸气或冷却水方法来控制；第三种方法是通过回流烃化液的温度进行调节。

烃化压力，烃化压力的考虑因素主要是在反应温度下苯的挥发度，在一个标准大气压下苯的沸点是80℃，而反应温度为95℃±5℃，因此必须维持一定的正压。通常反应压力为0.03～0.05MPa（表压）。

流量控制，鼓泡塔反应器在正常操作时，反应物苯在鼓泡塔中是连续相，乙烯是分散相。通常取苯的流量为乙烯流量的8～11倍，$AlCl_3$复合体加入量为苯流量的4%～5%。

捕集器，trap，用于气体采集的仪器。将一个能到－130℃以下的制冷盘管，放置在真空室中或油扩散泵的泵口，通过其表面的低温冷凝效应，迅速捕集真空系统的残余气体。

互动练习

19-1　What is the purpose of using a zeolite catalyst in the production of ethylbenzene from ethylene and benzene?

A）To reduce the reactor temperature and pressure

B）To increase the conversion and selectivity of the reaction

C）To improve the solvent flow rate

D）To reduce the number of by-products formed

19-2　How is the gas flow rate controlled in a bubble column reactor during normal operation?

A）By adjusting the reactor temperature

B）By changing the gas pressure

C）By adjusting the solvent flow rate

D）By adding more catalyst to the reactor

19-3　What is the first step in stopping a bubble column reactor?

A）Shut down the feed flow and pump

B）Regenerate the catalyst by burning off coke deposits

C）Cool down the reactor to a safe temperature

D）Remove the reaction product from the bottom of the reactor

19-4　What are some common operational issues in a hydrocarbon reactor?

A）Catalyst deactivation，pressure build-up，and cleaning of the reactor walls

B）Fouling of the reactor walls or internals，pressure build-up，and excessive solvent flow rate

C）Catalyst replacement，excessive coke deposition，and reactor temperature fluctuations

D）Catalyst deactivation，fouling of the reactor walls or internals，and pressure build-up due to excessive coke deposition

19-5　What should be done in the event of a reactor emergency?

A）Continue normal operation and monitor the situation

B）Shut down the reactor immediately and follow appropriate emergency procedures

C）Increase the gas flow rate to stabilize the reactor

D）Reduce the reactor temperature to prevent further damage

19.2 Common Troubleshooting and Maintenance Points for Bubble Column Reactors
鼓泡塔反应器常见故障处理与维护要点

Bubble column reactors are commonly used in the chemical industry for gas-liquid reactions. However，they are also prone to various types of malfunctions（运转失常）that can cause severe damage to the reactor and its surroundings. Common issues that can occur include deformation（变形）of the reactor body，cracking（破裂）of the reactor body，the tray crossing the stable operating region（稳定操作区），and the loss or corrosion of bubble elements.

In order to maintain the optimal performance of the reactor and prevent malfunctions，it is important to follow proper maintenance procedures. This includes conducting regular inspections during shutdown periods，inspecting the reactor body for signs of damage or corrosion，and checking the internals of the reactor for any signs of wear or corrosion.

The maintenance of the bubble elements is also crucial to ensure the proper operation of the reactor. Additionally，it is important to perform regular maintenance checks on the gas and liquid flow meters，as well as the other instrumentation used in the reactor. By following these maintenance procedures，operators can ensure the safe and efficient operation of the bubble column reactor.

技术理论

鼓泡塔反应器常见故障及处理方法见表19-1。

表 19-1　鼓泡塔反应器常见故障及处理方法

序号	故障现象	故障原因	处理方法
1	塔体出现变形	①塔局部腐蚀或过热使材料强度降低,而引起设备变形 ②开孔无补强或焊缝处的应力集中,使材料的内应力超过屈服极限而发生塑性变形 ③受外压设备,当工作压力超过临界工作压力时,设备失稳而变形	①防止局部腐蚀产生 ②矫正变形或切割下严重变形处,焊上补板 ③稳定正常操作
2	塔体出现裂缝	①局部变形加剧 ②焊接的内应力 ③封头过渡圆弧弯曲半径太小或未经返火便弯曲 ④水力冲击作用 ⑤结构材料缺陷 ⑥振动与温差的影响	裂缝修理
3	塔板越过稳定操作区	①气相负荷减小或增大,液相负荷减小 ②塔板不水平	①控制气相、液相流量。调整降液管、出入口堰高度 ②调整塔板水平度
4	鼓泡元件脱落和腐蚀掉	①安装不牢 ②操作条件破坏 ③泡罩材料不耐腐蚀	①重新调整 ②改善操作,加强管理 ③选择耐蚀材料,更新泡罩

鼓泡塔反应器维护要点如下。

（1）停车检查

塔设备停止生产时，要卸掉塔内压力，放出塔内所有存留物料，然后向塔内吹入蒸汽清洗。打开塔顶大盖（或塔顶气相出口）进行蒸煮、吹除、置换、降温，然后自上而下地打开塔体人孔。在检修前，要做好防火、防爆和防毒的安全措施，既要把塔内部的可燃性或有毒性介质彻底清洗吹净，又要对设备内及塔周围现场气体进行化验分析，达到安全检修的要求。

（2）塔体检查

① 每次检修都要检查各附件（压力表、安全阀与放空阀、温度计、单向阀、消防蒸汽阀等）是否灵活、准确。

② 检查塔体腐蚀、变形、壁厚减薄、裂纹及各部分焊接情况，进行超声波测厚和理化鉴定。并作详细记录，以备研究改进及作为下次检修的依据。经检查鉴定，如果认为对设计允许强度有影响时，可进行水压试验，其值参阅有关规定。

③ 检查塔内塔外内检污查垢和内部绝缘材料。

（3）塔内外检查

① 检查塔板各部件的结焦、污垢、堵塞情况，检查塔板、鼓泡构件和支承结构的腐蚀及变形情况。

② 检查塔板上各部件（出口堰、受液盘、降液管）的尺寸是否符合图纸及标准。

③ 对于浮阀塔板，应检查其浮阀的灵活性，是否有卡死、变形、冲蚀等现象，浮阀孔是否有堵塞。

④ 检查各种塔板、鼓泡构件等部件的紧固情况，是否有松动现象。

关键词详解

屈服极限，limit of yielding，也称流动极限。材料受外力到一定限度时，即使不增加负荷它仍继续发生明显的塑性变形。材料屈服极限是使试样产生给定的永久变形时所需要的应力，金属材料试样承受的外力超过材料的弹性极限时，虽然应力不再增加，但是试样仍发生明显的塑性变形，这种现象称为屈服，即材料承受外力到一定程度时，其变形不再与外力成正比而产生明显的塑性变形，产生屈服时的应力称为屈服极限。

塑性变形，plasticity，塑性变形是一种不可自行恢复的变形。工程材料及构件受载超过弹性变形范围之后将发生永久的变形，即卸除载荷后将出现不可恢复的变形，或称残余变形，这就是塑性变形。不是任何工程材料都具有塑性变形的能力。金属、塑料等都具有不同程度的塑性变形能力，故可称为塑性材料。

泡罩，blister，泡罩又称泡帽，是泡罩塔板的主要构件，有圆形泡罩及条形泡罩两类，主要采用圆形泡罩，它的外形如钟罩状。在泡罩的壁面开有很多齿缝，其形状有矩形、三角形及梯形。

泡罩内有升气管，在泡罩与升气管之间形成回转空间。

升气管固定在塔板上，气相通过升气管进入回转空间，以一定的喷出速度由齿缝喷出，与塔板上的液体形成鼓泡接触，进行传质过程。

安全阀，safety valve，安全阀是启闭件受外力作用下处于常闭状态，当设备或管道内的介质压力升高超过规定值时，通过向系统外排放介质来防止管道或设备内介质压力超过规定数值的特殊阀门。安全阀属于自动阀类，主要用于锅炉、压力容器和管道上，控制压力不超过规定值，对人身安全和设备运行起重要保护作用。注：安全阀必须经过压力试验才能使用。

浮阀，float valve，浮阀固定在容器壁上的所需高度上，并与供应管道连接。当液面升高时，浮球上升并通过杆臂来关闭阀。当液面下降时，供应管道打开，介质流入，直至再次达到各自的液位高度。

互动练习

19-6　What are some common malfunctions that can occur in a bubble column reactor?

A）Increased gas hold-up

B）Reduced mixing time

C）Tray crossing the stable operating region

D）Increased interfacial area

19-7　What is the importance of maintenance procedures in bubble column reactors?

A）To increase the interfacial area

B）To reduce the gas hold-up

C）To prevent malfunctions and damage to the reactor

D）To increase the mixing time

19-8　How can reactor operators prevent malfunctions in bubble column reactors?

A）By increasing the gas flow rate

B）By reducing the liquid flow rate

C）By following proper maintenance procedures

D）By increasing the reactor pressure

19-9　What is the purpose of inspecting the reactor body during maintenance procedures?

A）To increase the interfacial area

B）To reduce the gas hold-up

C）To check for signs of damage or corrosion

D）To increase the mixing time

19-10　Why is the maintenance of bubble elements crucial for proper operation of the reactor?

A）To increase the interfacial area

B）To reduce the gas hold-up

C）To ensure proper mixing of gas and liquid

D）To increase the reactor pressure

任务20

Operation and Control of Packed Tower Reactors
填料塔反应器操作与控制

任务要点

    本任务介绍填料塔反应器的操作与控制技术，读者将学习如何通过调节气液比、优化流体分布等方式，提高设备运行效率，降低能耗并确保设备稳定运行。

学习目标

知识目标

（1）掌握填料塔反应器操作中的动态行为及影响因素。

（2）了解填料塔反应器在操作中可能出现的问题，如液泛和干塔现象。

技能目标

（1）能优化填料塔反应器的运行条件以减少能耗和提高效率。

（2）能诊断并解决操作中的常见故障。

价值目标

（1）强调环保生产和资源节约的重要性。

（2）在设备运行中贯彻持续改进理念。

20.1 **Operation and Control of Carbon Dioxide Contact Absorption Towers**
二氧化碳接触塔操作与控制

In the process of producing ethylene oxide，a CO_2 contact tower and a packed tower are used for gas purification. The process flow begins with the removal of impurities and moisture（水分）from the feed gas，which is then cooled and compressed before entering the CO_2 contact tower for removal of CO_2. The gas then enters the packed tower，where it is further purified by removing organic impurities using special adsorption materials.

Before starting the operation，it is important to prepare the towers by mechanically cleaning and inspecting them，washing with water and alkali，and adding carbonate. The gas circulation through the carbonate system ensures the dry operation of the towers.

During normal operation，it is important to monitor the pressure，temperature，and flow rate of the gas in the towers. The packed tower should be periodically inspected for any signs of clogging or damage to the packing material. In the event of any issues，the gas flow rate or temperature should be adjusted，or the packing material should be replaced.

To stop the operation，the gas flow should be gradually reduced to prevent sudden pressure changes. The towers should be thoroughly washed with water and alkali，and the packing material should be inspected and replaced if necessary.

Common issues in the CO_2 contact tower include high pressure drop and loss of CO_2 adsorption capacity. These issues can be resolved by adjusting the gas flow rate or replacing the adsorption material. In the packed tower，clogging or damage to the packing material can lead to reduced purification efficiency，which can be remedied by replacing the packing material.

技术理论

CO_2 吸收工艺流程如图 20-1 所示。来自循环压缩机出口循环气（含 CO_2 体积分数为 8.1%）与回收的压缩机出口气体汇合后（含 CO_2 体积分数大约为 8.9%），这股富二氧化碳循环气进到预饱和器部分，在预饱和器内循环气同来自接触塔分离罐的洗涤水逆流接触，直接进行热交换，使循环气温度升高。然后，富二氧化碳循环气进入接触塔的底部，在此循环气与贫碳酸钾溶液接触，循环气中的碳酸钾转化为碳酸氢钾，使循环气中的二氧化碳含量减少到 CO_2 体积分数为 3.86%，贫二氧化碳循环气从接触塔的顶部流到分离罐底部。在接触塔分离罐内，贫二氧化碳循环气同来自洗涤水冷却器的水直接接触，被冷却和洗涤。洗涤后的贫二氧化碳循环气离开预饱和器和循环气分离罐，流到塔底部的分离罐，离开分离罐的贫二氧化碳循环气返回到反应单元。来自接触塔底部的富二氧化碳碳酸盐溶液减压进入再生塔进料闪蒸罐。在此，溶解在富碳酸盐溶液中的所有碳氢化合物基本上都闪蒸出来，进入气相，作为塔顶采出物，并同再吸收塔塔顶气体一起经回收压缩机送回预饱和罐。

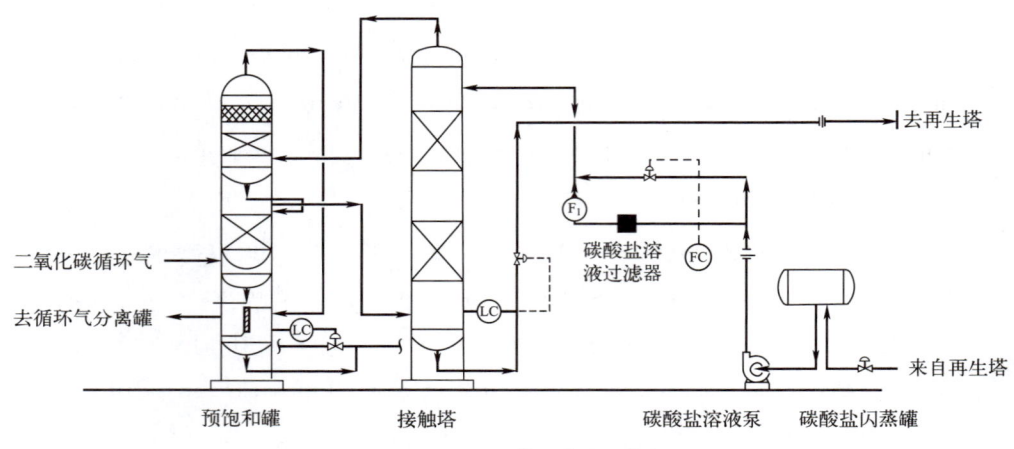

二氧化碳循环气 →

去循环气分离罐 →

碳酸盐溶液过滤器

去再生塔

来自再生塔

预饱和罐　　　接触塔　　　碳酸盐溶液泵　碳酸盐闪蒸罐

图 20-1　CO₂ 吸收工艺流程简图

关键词详解

　　开车前准备，preparation before start up，二氧化碳脱除系统的准备和试验主要包括机械清洗及检查、水洗、碱洗、加入碳酸盐和通过碳酸盐系统的循环气体干运转几个步骤。

　　正常开车，normal startup，正常开车程序包括：① 初次开车时，循环气系统必须用氮气升压；② 循环水系统必须运行；③ 循环气压缩机运行，小股物流经过反应器，大部分物流经过旁路；④ 反应器及反应器冷却器蒸汽发生系统运行，反应温度为 200℃；⑤ 二氧化碳脱除系统处于运行状态，有一小股循环气通过；⑥ 所有气体分析器投入运行并已标定。

　　正常停车，normal shutdown，正常停车程序包括：① 当所有的氧气都已耗尽，不再有二氧化碳生成时，切断二氧化碳接触塔气体进出口。② 在设计浓度下，碳酸氢盐将在 55% 时析出。在碳酸氢盐全部转化之前，碳酸盐再生必须继续进行。当溶液完全再生后，停止向再生塔通蒸汽。③ 如果长期停车（5 天或更长时间），应该停碳酸盐溶液泵，塔内的液体排到碳酸盐贮罐。碳酸盐贮罐的温度必须保持在 70℃，因碳酸盐系统重新启动要比环氧乙烷生产装置其他部分开车提前很久。

　　正常操作，normal operation，二氧化碳接触塔的控制参数为碳酸盐流量、液位和气体流量。

（1）碳酸盐流量

　　碳酸盐溶液所需的流量应保持在设计值。如果反应器入口二氧化碳浓度连续超过设计值（或接触塔出口二氧化碳超过设计值），需少量增加碳酸盐溶液流量，应检查贫碳酸盐溶液中碳酸氢盐/碳酸盐的浓度。

（2）消泡

接触塔发泡的结果非常有害，会使碳酸盐进入反应系统。使用一种合适的消泡剂来控制可能发生的起泡。消泡剂应少量添加，不能大量一次加入。初期每天加入约 25～50mL。消泡剂的加入数量和频率应根据循环气通过二氧化碳系统的压差来调节。定期控制碳酸盐溶液的起泡。

（3）接触塔液位

一般来说，接触塔液位最好保持在液位控制器量程的 50%（大约）。

（4）循环气流量

通过接触塔的循环气流量由流量控制阀控制，它控制流经二氧化碳脱除系统的循环气流量，位于从接触塔分离罐到循环气分离罐上游的循环气系统的出口管线上。脱除反应器中产生的二氧化碳，并在稳定的状态下保持循环气中的二氧化碳浓度。通过调节循环压缩机出口的阀门来间接控制通过接触塔的循环气流量，这个流量应调节到维持在反应器进口循环气中的二氧化碳在 7%（摩尔分数）或低于 7%。

当进入接触塔的气体流量变化时，应注意防止起泡或液泛。

（5）洗涤水流量

洗涤水被送到接触塔洗涤部分，然后在液位控制下进入接触塔的预饱和罐部分。在洗涤部分，任何由贫二氧化碳循环气带来的碳酸盐都被洗涤下来。在预饱和罐部分，富二氧化碳循环气被加热后进入接触塔。避免过大的流量，以防止可能在液体分布器上发生起泡并使水进入反应系统。

（6）预饱和罐液位

一般来说，预饱和罐液位应保持在控制器量程的 50% 左右。

（7）碳酸盐溶液浓度

贫碳酸盐溶液应维持在当量碳酸钾在 25%（质量分数）。较低的浓度将减少二氧化碳脱除系统的能力。较高的浓度可能使重碳酸盐沉淀。

互动练习

20-1　What is the purpose of the carbon dioxide contact tower in the production process of epoxy ethane?

　A）To remove carbon dioxide from the reaction mixture

　B）To add carbon dioxide to the reaction mixture

　C）To filter impurities from the reaction mixture

　D）To maintain the temperature of the reaction mixture

20-2　What is the initial preparation required before starting the carbon dioxide contact tower in the production process of epoxy ethane?

　A）Mechanical cleaning and inspection

　B）Acid washing of the tower

　C）Adding carbon dioxide to the reaction mixture

　D）Flushing the tower with water

20-3　What is the purpose of adding carbonates before starting the carbon dioxide contact tower in the production process of epoxy ethane?

　A）To neutralize any acidic impurities in the tower

　B）To enhance the contact between carbon dioxide and the reaction mixture

　C）To increase the temperature of the reaction mixture

　D）To prevent corrosion of the tower

20-4　How can the carbon dioxide contact tower be operated during normal operation?

　A）By recycling the gas through the carbonate system

　B）By adding more carbon dioxide to the reaction mixture

　C）By increasing the temperature of the reaction mixture

　D）By reducing the pressure inside the tower

20-5　What is the common abnormal phenomenon that can occur in the carbon dioxide contact tower during operation, and how can it be handled?

　A）Clogging of the tower, which can be cleaned with acid washing

　B）Corrosion of the tower, which can be prevented by adding inhibitors to the reaction mixture

　C）Leakage of the tower, which can be repaired with welding

　D）Poor gas-liquid contact, which can be improved by adjusting the gas flow rate or increasing the contact time

20.2 Common Troubleshooting for Packed Tower Reactors
填料塔反应器常见故障处理与维护要点

Packed tower reactors are widely used in the chemical industry due to their high efficiency and compact size. However，like any other equipment，they are prone to breakdowns（出故障）and require regular maintenance to ensure optimal performance.

Common problems in packed bed reactors include clogging，fouling，and bed compaction（基床压实）. Clogging can be caused by the accumulation of solids or the formation of a viscous layer on the packing surface. Fouling occurs when reactants or products deposit on the packing，reducing its effectiveness （有效性）. Bed compaction is a gradual loss of void space between the packing materials，which can lead to channeling or uneven flow distribution. To prevent these problems，regular maintenance is essential，including inspection，cleaning，and replacement of damaged or worn-out components.

In addition，proper operation，and control，including flow rate，temperature，and pressure monitoring，can help prevent problems before they occur. The use of high-quality packing materials and the selection of appropriate operating conditions can also reduce the risk of equipment failure.

技术理论

填料塔反应器生产过程中常见故障与处理方法见表 20-1。

表 20-1　填料塔反应器生产过程中常见故障与处理方法

序号	异常现象	原因分析判断	操作处理方法
1	①解吸塔塔顶冷却水中断 ②解吸塔塔顶温度和压力升高,入口阀处于常开状态,冷却水流量为零	冷却水中断	①打开调压阀保压,关闭加热蒸汽阀门,停用再沸器 ②停止向吸收塔进富 CO_2 循环气 ③停止向解吸塔进料 ④关闭循环气出口阀 ⑤停止向吸收塔加入碳酸盐溶液,停止解吸塔回流 ⑥事故解除后按热状态开车操作

续表

序号	异常现象	原因分析判断	操作处理方法
2	仪表风中断	各调节阀全开或全闭	①打开并调节吸收塔碳酸盐溶液流量调节阀的旁通阀，并使流量维持在正常值 ②打开并调节吸收塔塔釜溶液流量调节阀的旁通阀，并使流量维持在正常值 ③打开吸收塔温度和压力调节阀的旁通阀，并使之维持在正常值 ④打开并调节控制解吸塔的液位和回流量调节阀的旁通阀，并使流量维持在正常值
3	循环气中 CO_2 浓度偏高	①进氧量偏大 ②输送碳酸盐溶液管线堵塞	①设法在系统内碳酸盐溶液倒空之前使碳酸盐溶液流动，以防管线内结晶堵塞 ②立即停止氧气和乙烯进料 ③着手解决碳酸盐溶液倒空及洗塔问题
4	吸收塔有较高液位	再吸收塔釜液泵故障	①将备用泵投入使用 ②如果备用泵不能使用，反应系统应紧急停车
5	①再生塔中液位升高 ②吸收塔顶温度升高，压力上升	碳酸盐溶液泵故障	①将备用泵投入使用 ②如果备用泵不能投入使用，应紧急停车，要求停车在由于 CO_2 的积累造成循环气体压力过于升高之前进行

填料塔维护要点如下。

① 定期检查、清理、更换莲蓬头或溢流管，保持不堵塞、不破损、不偏斜，使喷淋装置能把液体均匀地分布到填料上；

② 进塔气体的压力和流速不能过大，否则将带走填料或使其紊乱，严重降低气液两相接触效率；

③ 控制进气温度，防止塑料填料软化或变质，增加气流阻力；

④ 进塔的液体不能含有杂物，太脏时应过滤，避免杂物堵塞填料缝隙；

⑤ 定期检查、防腐、清理塔壁，防止腐蚀、冲刷、挂疤等缺陷；

⑥ 定期检查箅板腐蚀程度，如果腐蚀变薄则应更新，防止脱落；

⑦ 定期测量塔壁厚度并观察塔体有无渗漏，发现后及时修补；

⑧ 经常检查液面，不要淹没气体进口，防止引起振动和异常响声；

⑨ 经常观察基础下沉情况，注意塔体有无倾斜；

⑩ 保持塔体油漆完整，外观无挂疤，清洁卫生；

⑪ 定期打开排污阀门，排放塔底积存脏物和碎填料；

⑫ 冬季停用时，应将液体放净，防止冻结；

⑬ 如果压力突然下降，可能的原因是发生了泄漏。如果压力上升，可能的

原因是填料阻力增加或设备管道堵塞；

⑭ 防腐层和保温层损坏，此时要对室外保温的设备进行检查，着重检查温度在100℃以下的雨水浸入处、保温材料变质处、长期经外来微量腐蚀性流体侵蚀处。

关键词详解

应力腐蚀，stress corrosion，是指在拉应力作用下，金属在腐蚀介质中引起的破坏。这种腐蚀一般均穿过晶粒，即所谓穿晶腐蚀。应力腐蚀是由残余或外加应力导致的应变和腐蚀联合作用产生的材料破坏过程。应力腐蚀导致材料的断裂称为应力腐蚀断裂。

互动练习

20-6　What is the purpose of the fillers in a packed tower reactor?

A）To improve heat transfer efficiency

B）To increase the pressure，drop across the bed

C）To decrease the reaction rate

D）To promote mass transfer between the two phases

20-7　What are some common maintenance tasks for a packed tower reactor?

A）Cleaning the feed lines

B）Checking for leaks in the reactor shell

C）Replacing damaged fillers

D）All of the above

20-8　What is a potential consequence of the fillers becoming clogged in a packed tower reactor?

A）Reduced mass transfer efficiency

B）Increased reaction rate

C）Decreased pressure drops across the bed

D）None of the above

20-9　Which of the following is a common problem with packed tower reactors?

A）Reactor shell deformation

B）Plate crossover in the column

C）Loss of filler material

D）All of the above

20-10　How can the fillers be selected to improve reactor performance in a packed tower reactor?

A）By increasing the surface area of the fillers

B）By using a high-density filler material

C）By decreasing the size of the fillers

D）All of the above

参考文献

[1] 陈炳和，许宁. 化学反应过程与设备. 4 版. 北京：化学工业出版社，2020.

[2] 李绍芬. 反应工程（英文版）. 北京：化学工业出版社，2019.

[3] 朱炳辰. 化学反应工程. 5 版. 北京：化学工业出版社，2020.

[4] 陈建峰，陈甘棠. 化学反应工程. 5 版. 北京：化学工业出版社，2024.

[5] 陆敏. 化学制药工业与反应器. 5 版. 北京：化学工业出版社，2023.

[6] 许志美. 化学反应工程. 北京：化学工业出版社，2019.

[7] 陈敏恒. 化工原理（少学时）. 3 版. 上海：华东理工大学出版社，2019.

[8] Fogler H. S. Elements of Chemical Reaction Engineering. 6th ed. London：Pearson，2020.

[9] Fogler H. S. Essentials of Chemical Reaction Engineering. 2nd ed. London：Pearson，2017.

[10] Levenspiel O. Chemical Reaction Engineering. 3rd ed. 北京：化学工业出版社，2002.

[11] Smith J. M.，Van Ness H. C.，Abbott M. M.，Swihart M. T. Introduction to Chemical Engineering Thermodynamics. 8th ed. New York：McGraw-Hill Education，2018.

[12] Green D. W.，Southard M. Z. Perry's Chemical Engineers' Handbook，9th ed. New York：McGraw-Hill Education，2019.

[13] Hill C. G.；Root，T. W. Introduction to Chemical Engineering Kinetics and Reactor Design，2nd ed. New Jersey：Wiley，2014.

[14] 中石化上海工程有限公司. 化工工艺设计手册（下册）. 5 版. 北京：化学工业出版社，2019.

[15] 张卫红，李为民. 化学反应工程. 3 版. 北京：中国石化出版社，2020.

[16] 王尚弟. 催化及工程导论. 3 版. 北京：化学工业出版社，2015.

[17] 袁渭康，王静康，费维扬，欧阳平凯. 化学工程手册. 3 版. 北京：化学工业出版社，2019.

[18] 单国荣，杜淼，朱利平. 聚合反应工程基础. 2 版. 北京：化学工业出版社，2021.

[19] 樊亚娟，薛叙明. 化工仿真操作实训. 北京：化学工业出版社，2023.